I0516679

Tucholsky Wagner Zola Scott Sydow Freud Schlegel
Turgenev Wallace Fonatne
Twain Walther von der Vogelweide Fouqué Friedrich II. von Preußen
Weber Freiligrath Frey
Kant Ernst
Fechner Fichte Weiße Rose von Fallersleben Richthofen Frommel
Hölderlin
Engels Fielding Tacitus Dumas
Fehrs Faber Flaubert Eichendorff
Maximilian I. von Habsburg Eliasberg Ebner Eschenbach
Feuerbach Fock Zweig
Ewald Eliot Vergil
Goethe Elisabeth von Österreich London
Mendelssohn Balzac Shakespeare Dostojewski Ganghofer
Lichtenberg Rathenau Doyle Gjellerup
Trackl Stevenson Hambruch
Mommsen Tolstoi Lenz Droste-Hülshoff
Thoma Hanrieder
Dach von Arnim Hägele Hauff Humboldt
Reuter Verne
Karrillon Rousseau Hagen Hauptmann Gautier
Garschin
Damaschke Defoe Hebbel Baudelaire
Descartes
Wolfram von Eschenbach Schopenhauer Hegel Kussmaul Herder
Bronner Darwin Dickens Rilke George
Melville Grimm Jerome Bebel
Campe Horváth Aristoteles Proust
Bismarck Vigny Barlach Voltaire Federer Herodot
Gengenbach Heine
Storm Casanova Tersteegen Gilm Grillparzer Georgy
Chamberlain Lessing Langbein Gryphius
Brentano Lafontaine
Strachwitz Claudius Schiller Kralik Iffland Sokrates
Katharina II. von Rußland Bellamy Schilling
Gerstäcker Raabe Gibbon Tschechow
Löns Hesse Hoffmann Gogol Wilde Gleim Vulpius
Luther Heym Hofmannsthal Klee Hölty Morgenstern Goedicke
Roth Heyse Klopstock Kleist
Luxemburg Puschkin Homer Mörike Musil
La Roche Horaz
Machiavelli Kierkegaard Kraft Kraus
Navarra Aurel Musset Kind Moltke
Nestroy Marie de France Lamprecht Kirchhoff Hugo
Laotse Ipsen Liebknecht
Nietzsche Nansen Ringelnatz
Marx Lassalle Gorki Klett Leibniz
von Ossietzky May vom Stein Lawrence Irving
Petalozzi Knigge
Platon Pückler Michelangelo Kock Kafka
Sachs Poe Liebermann Korolenko
de Sade Praetorius Mistral Zetkin

The publishing house tredition has created the series **TREDITION CLASSICS**. It contains classical literature works from over two thousand years. Most of these titles have been out of print and off the bookstore shelves for decades.

The book series is intended to preserve the cultural legacy and to promote the timeless works of classical literature. As a reader of a **TREDITION CLASSICS** book, the reader supports the mission to save many of the amazing works of world literature from oblivion.

The symbol of **TREDITION CLASSICS** is Johannes Gutenberg (1400 – 1468), the inventor of movable type printing.

With the series, tredition intends to make thousands of international literature classics available in printed format again – worldwide.

All books are available at book retailers worldwide in paperback and in hardcover. For more information please visit: www.tredition.com

tredition was established in 2006 by Sandra Latusseck and Soenke Schulz. Based in Hamburg, Germany, tredition offers publishing solutions to authors and publishing houses, combined with worldwide distribution of printed and digital book content. tredition is uniquely positioned to enable authors and publishing houses to create books on their own terms and without conventional manufacturing risks.

For more information please visit: www.tredition.com

Oxy-Acetylene Welding and Cutting Electric, Forge and Thermit Welding together with related methods and materials used in metal working and the oxygen process for removal of carbon

Harold P. (Harold Phillips) Manly

Imprint

This book is part of the TREDITION CLASSICS series.

Author: Harold P. (Harold Phillips) Manly
Cover design: toepferschumann, Berlin (Germany)

Publisher: tredition GmbH, Hamburg (Germany)
ISBN: 978-3-8495-0900-2

www.tredition.com
www.tredition.de

Copyright:
The content of this book is sourced from the public domain.

The intention of the TREDITION CLASSICS series is to make world literature in the public domain available in printed format. Literary enthusiasts and organizations worldwide have scanned and digitally edited the original texts. tredition has subsequently formatted and redesigned the content into a modern reading layout. Therefore, we cannot guarantee the exact reproduction of the original format of a particular historic edition. Please also note that no modifications have been made to the spelling, therefore it may differ from the orthography used today.

PREFACE

In the preparation of this work, the object has been to cover not only the several processes of welding, but also those other processes which are so closely allied in method and results as to make them a part of the whole subject of joining metal to metal with the aid of heat.

The workman who wishes to handle his trade from start to finish finds that it is necessary to become familiar with certain other operations which precede or follow the actual joining of the metal parts, the purpose of these operations being to add or retain certain desirable qualities in the materials being handled. For this reason the following subjects have been included: Annealing, tempering, hardening, heat treatment and the restoration of steel.

In order that the user may understand the underlying principles and the materials employed in this work, much practical information is given on the uses and characteristics of the various metals; on the production, handling and use of the gases and other materials which are a part of the equipment; and on the tools and accessories for the production and handling of these materials.

An examination will show that the greatest usefulness of this book lies in the fact that all necessary information and data has been included in one volume, making it possible for the workman to use one source for securing a knowledge of both principle and practice, preparation and finishing of the work, and both large and small repair work as well as manufacturing methods used in metal working.

An effort has been made to eliminate all matter which is not of direct usefulness in practical work, while including all that those engaged in this trade find necessary. To this end, the descriptions have been limited to those methods and accessories which are found in actual use today. For the same reason, the work includes the application of the rules laid down by the insurance underwriters which govern this work as well as instructions for the proper care and handling of the generators, torches and materials found in the shop.

Special attention has been given to definite directions for handling the different metals and alloys which must be handled. The instructions have been arranged to form rules which are placed in the order of their use during the work described and the work has been subdivided in such a way that it will be found possible to secure information on any one point desired without the necessity of spending time in other fields.

The facts which the expert welder and metalworker finds it most necessary to have readily available have been secured, and prepared especially for this work, and those of most general use have been combined with the chapter on welding practice to which they apply.

The size of this volume has been kept as small as possible, but an examination of the alphabetical index will show that the range of subjects and details covered is complete in all respects. This has been accomplished through careful classification of the contents and the elimination of all repetition and all theoretical, historical and similar matter that is not absolutely necessary.

Free use has been made of the information given by those manufacturers who are recognized as the leaders in their respective fields, thus insuring that the work is thoroughly practical and that it represents present day methods and practice.

THE AUTHOR.

CONTENTS

CHAPTER I
METALS AND ALLOYS—HEAT TREATMENT:—The Use and Characteristics of the
Industrial Alloys and Metal Elements—Annealing, Hardening, Tempering and
Case Hardening of Steel

CHAPTER II
WELDING MATERIALS:—Production, Handling and Use of the Gases, Oxygen and
Acetylene—Welding Rods—Fluxes—Supplies and Fixtures

CHAPTER III
ACETYLENE GENERATORS:—Generator Requirements and Types—Construction—Care and Operation of Generators.

CHAPTER IV
WELDING INSTRUMENTS:—Tank and Regulating Valves and Gauges—High, Low and
Medium Pressure Torches—Cutting Torches—Acetylene-Air Torches

CHAPTER V
OXY-ACETYLENE WELDING PRACTICE:—Preparation of Work—Torch Practice—
Control of the Flame—Welding Various Metals and Alloys—Tables of
Information Required in Welding Operations

CHAPTER VI

ELECTRIC WELDING: — Resistance Method — Butt, Spot and Lap Welding — Troubles and Remedies — Electric Arc Welding

CHAPTER VII

HAND FORGING AND WELDING: — Blacksmithing, Forging and Bending — Forge
Welding Methods

CHAPTER VIII

SOLDERING, BRAZING AND THERMIT WELDING: — Soldering Materials and Practice —
Brazing — Thermit Welding

CHAPTER IX

OXYGEN PROCESS FOR REMOVAL OF CARBON

INDEX

OXY-ACETYLENE WELDING AND CUTTING, ELECTRIC AND THERMIT WELDING

CHAPTER I

METALS AND THEIR ALLOYS—HEAT TREATMENT

THE METALS

Iron.—Iron, in its pure state, is a soft, white, easily worked metal. It is the most important of all the metallic elements, and is, next to aluminum, the commonest metal found in the earth.

Mechanically speaking, we have three kinds of iron: wrought iron, cast iron and steel. Wrought iron is very nearly pure iron; cast iron contains carbon and silicon, also chemical impurities; and steel contains a definite proportion of carbon, but in smaller quantities than cast iron.

Pure iron is never obtained commercially, the metal always being mixed with various proportions of carbon, silicon, sulphur, phosphorus, and other elements, making it more or less suitable for different purposes. Iron is magnetic to the extent that it is attracted by magnets, but it does not retain magnetism itself, as does steel. Iron forms, with other elements, many important combinations, such as its alloys, oxides, and sulphates.

[Illustration: Figure 1.—Section Through a Blast Furnace]

Cast Iron.—Metallic iron is separated from iron ore in the blast furnace (Figure 1), and when allowed to run into moulds is called cast iron. This form is used for engine cylinders and pistons, for brackets, covers, housings and at any point where its brittleness is not objectionable. Good cast iron breaks with a gray fracture, is free from blowholes or roughness, and is easily machined, drilled, etc. Cast iron is slightly lighter than steel, melts at about 2,400 degrees in practice, is about one-eighth as good an electrical conductor as copper and has a tensile strength of 13,000 to 30,000 pounds per square inch. Its compressive strength, or resistance to crushing, is very great. It has excellent wearing qualities and is not easily warped and deformed by heat. Chilled iron is cast into a metal mould so that the outside is cooled quickly, making the surface very hard and difficult to cut and giving great resistance to wear. It is used for making cheap gear wheels and parts that must withstand surface friction.

Malleable Cast Iron. — This is often called simply malleable iron. It is a form of cast iron obtained by removing much of the carbon from cast iron, making it softer and less brittle. It has a tensile strength of 25,000 to 45,000 pounds per square inch, is easily machined, will stand a small amount of bending at a low red heat and is used chiefly in making brackets, fittings and supports where low cost is of considerable importance. It is often used in cheap constructions in place of steel forgings. The greatest strength of a malleable casting, like a steel forging, is in the surface, therefore but little machining should be done.

Wrought Iron. — This grade is made by treating the cast iron to remove almost all of the carbon, silicon, phosphorus, sulphur, manganese and other impurities. This process leaves a small amount of the slag from the ore mixed with the wrought iron.

Wrought iron is used for making bars to be machined into various parts. If drawn through the rolls at the mill once, while being made, it is called "muck bar;" if rolled twice, it is called "merchant bar" (the commonest kind), and a still better grade is made by rolling a third time. Wrought iron is being gradually replaced in use by mild rolled steels.

Wrought iron is slightly heavier than cast iron, is a much better electrical conductor than either cast iron or steel, has a tensile strength of 40,000 to 60,000 pounds per square inch and costs slightly more than steel. Unlike either steel or cast iron, wrought iron does not harden when cooled suddenly from a red heat.

Grades of Irons.—The mechanical properties of cast iron differ greatly according to the amount of other materials it contains. The most important of these contained elements is carbon, which is present to a degree varying from 2 to 5-1/2 per cent. When iron containing much carbon is quickly cooled and then broken, the fracture is nearly white in color and the metal is found to be hard and brittle. When the iron is slowly cooled and then broken the fracture is gray and the iron is more malleable and less brittle. If cast iron contains sulphur or phosphorus, it will show a white fracture regardless of the rapidity of cooling, being brittle and less desirable for general work.

Steel.—Steel is composed of extremely minute particles of iron and carbon, forming a network of layers and bands. This carbon is a smaller proportion of the metal than found in cast iron, the percentage being from 3/10 to 2-1/2 per cent.

Carbon steel is specified according to the number of "points" of carbon, a point being one one-hundredth of one per cent of the weight of the steel. Steel may contain anywhere from 30 to 250 points, which is equivalent to saying, anywhere from 3/10 to 2-1/2 per cent, as above. A 70-point steel would contain 70/100 of one per cent or 7/10 of one per cent of carbon by weight. The percentage of carbon determines the hardness of the steel, also many other qualities, and its suitability for various kinds of work. The more carbon contained in the steel, the harder the metal will be, and, of course, its brittleness increases with the hardness. The smaller the grains or particles of iron which are separated by the carbon, the stronger the steel will be, and the control of the size of these particles is the object of the science of heat treatment.

In addition to the carbon, steel may contain the following:

Silicon, which increases the hardness, brittleness, strength and difficulty of working if from 2 to 3 per cent is present.

Phosphorus, which hardens and weakens the metal but makes it easier to cast. Three-tenths per cent of phosphorus serves as a hardening agent and may be present in good steel if the percentage of carbon is low. More than this weakens the metal.

Sulphur, which tends to make the metal hard and filled with small holes.

Manganese, which makes the steel so hard and tough that it can with difficulty be cut with steel tools. Its hardness is not lessened by annealing, and it has great tensile strength.

Alloy steel has a varying but small percentage of other elements mixed with it to give certain desired qualities. Silicon steel and manganese steel are sometimes classed as alloy steels. This subject is taken up in the latter part of this chapter under *Alloys*, where the various combinations and their characteristics are given consideration.

Steel has a tensile strength varying from 50,000 to 300,000 pounds per square inch, depending on the carbon percentage and the other alloys present, as well as upon the texture of the grain. Steel is heavier than cast iron and weighs about the same as wrought iron. It is about one-ninth as good a conductor of electricity as copper.

Steel is made from cast iron by three principal processes: the crucible,
Bessemer and open hearth.

Crucible steel is made by placing pieces of iron in a clay or graphite crucible, mixed with charcoal and a small amount of any desired alloy. The crucible is then heated with coal, oil or gas fires until the iron melts, and, by absorbing the desired elements and giving up or changing its percentage of carbon, becomes steel. The molten steel is then poured from the crucible into moulds or bars for use. Crucible steel may also be made by placing crude steel in the crucibles in place of the iron. This last method gives the finest grade of metal and the crucible process in general gives the best grades of steel for mechanical use.

[Illustration: Figure 2. — A Bessemer Converter]

Bessemer steel is made by heating iron until all the undesirable elements are burned out by air blasts which furnish the necessary oxygen. The iron is placed in a large retort called a converter, being poured, while at a melting heat, directly from the blast furnace into the converter. While the iron in the converter is molten, blasts of air are forced through the liquid, making it still hotter and burning out the impurities together with the carbon and manganese. These two elements are then restored to the iron by adding spiegeleisen (an alloy of iron, carbon and manganese). A converter holds from 5 to 25 tons of metal and requires about 20 minutes to finish a charge. This makes the cheapest steel.

[Illustration: Figure 3. — An Open Hearth Furnace]

Open hearth steel is made by placing the molten iron in a receptacle while currents of air pass over it, this air having itself been highly heated by just passing over white hot brick (Figure. 3). Open hearth steel is considered more uniform and reliable than Bessemer, and is used for springs, bar steel, tool steel, steel plates, etc.

Aluminum is one of the commonest industrial metals. It is used for gear cases, engine crank cases, covers, fittings, and wherever lightness and moderate strength are desirable.

Aluminum is about one-third the weight of iron and about the same weight as glass and porcelain; it is a good electrical conductor (about one-half as good as copper); is fairly strong itself and gives great strength to other metals when alloyed with them. One of the greatest advantages of aluminum is that it will not rust or corrode under ordinary conditions. The granular formation of aluminum makes its strength very unreliable and it is too soft to resist wear.

Copper is one of the most important metals used in the trades, and the best commercial conductor of electricity, being exceeded in this respect only by silver, which is but slightly better. Copper is very malleable and ductile when cold, and in this state may be easily worked under the hammer. Working in this way makes the copper stronger and harder, but less ductile. Copper is not affected by air, but acids cause the formation of a green deposit called verdigris.

Copper is one of the best conductors of heat, as well as electricity, being used for kettles, boilers, stills and wherever this quality is desirable. Copper is also used in alloys with other metals, forming an important part of brass, bronze, german silver, bell metal and gun metal. It is about one-eighth heavier than steel and has a tensile strength of about 25,000 to 50,000 pounds per square inch.

Lead.—The peculiar properties of lead, and especially its quality of showing but little action or chemical change in the presence of other elements, makes it valuable under certain conditions of use. Its principal use is in pipes for water and gas, coverings for roofs and linings for vats and tanks. It is also used to coat sheet iron for similar uses and as an important part of ordinary solder.

Lead is the softest and weakest of all the commercial metals, being very pliable and inelastic. It should be remembered that lead and all its compounds are poisonous when received into the system. Lead is more than one-third heavier than steel, has a tensile strength of only about 2,000 pounds per square inch, and is only about one-tenth as good a conductor of electricity as copper.

Zinc.—This is a bluish-white metal of crystalline form. It is brittle at ordinary temperatures and becomes malleable at about 250 to 300 degrees Fahrenheit, but beyond this point becomes even more brittle than at ordinary temperatures. Zinc is practically unaffected by air or moisture through becoming covered with one of its own compounds which immediately resists further action. Zinc melts at low temperatures, and when heated beyond the melting point gives off very poisonous fumes.

The principal use of zinc is as an alloy with other metals to form brass, bronze, german silver and bearing metals. It is also used to cover the surface of steel and iron plates, the plates being then called galvanized.

Zinc weighs slightly less than steel, has a tensile strength of 5,000 pounds per square inch, and is not quite half as good as copper in conducting electricity.

Tin resembles silver in color and luster. Tin is ductile and malleable and slightly crystalline in form, almost as heavy as steel, and has a tensile strength of 4,500 pounds per square inch.

The principal use of tin is for protective platings on household utensils and in wrappings of tin-foil. Tin forms an important part of many alloys such as babbitt, Britannia metal, bronze, gun metal and bearing metals.

Nickel is important in mechanics because of its combinations with other metals as alloys. Pure nickel is grayish-white, malleable, ductile and tenacious. It weighs almost as much as steel and, next to manganese, is the hardest of metals. Nickel is one of the three magnetic metals, the others being iron and cobalt. The commonest alloy containing nickel is german silver, although one of its most important alloys is found in nickel steel. Nickel is about ten per cent heavier than steel, and has a tensile strength of 90,000 pounds per square inch.

Platinum.—This metal is valuable for two reasons: it is not affected by the air or moisture or any ordinary acid or salt, and in addition to this property it melts only at the highest temperatures. It is a fairly good electrical conductor, being better than iron or steel. It is nearly three times as heavy as steel and its tensile strength is 25,000 pounds per square inch.

ALLOYS

An alloy is formed by the union of a metal with some other material, either metal or non-metallic, this union being composed of two or more elements and usually brought about by heating the substances together until they melt and unite. Metals are alloyed with materials which have been found to give to the metal certain characteristics which are desired according to the use the metal will be put to.

The alloys of metals are, almost without exception, more important from an industrial standpoint than the metals themselves. There are innumerable possible combinations, the most useful of which are here classed under the head of the principal metal entering into their composition.

Steel.—Steel may be alloyed with almost any of the metals or elements, the combinations that have proven valuable numbering more than a score. The principal ones are given in alphabetical order, as follows:

Aluminum is added to steel in very small amounts for the purpose of preventing blow holes in castings.

Boron increases the density and toughness of the metal.

Bronze, added by alloying copper, tin and iron, is used for gun metal.

Carbon has already been considered under the head of steel in the section devoted to the metals. Carbon, while increasing the strength and hardness, decreases the ease of forging and bending and decreases the magnetism and electrical conductivity. High carbon steel can be welded only with difficulty. When the percentage of carbon is low, the steel is called "low carbon" or "mild" steel. This is used for rods and shafts, and called "machine" steel. When the carbon percentage is high, the steel is called "high carbon" steel, and it is used in the shop as tool steel. One-tenth per cent of carbon gives steel a tensile strength of 50,000 to 65,000 pounds per square inch; two-tenths per cent gives from 60,000 to 80,000; four-tenths per cent gives 70,000 to 100,000, and six-tenths per cent gives 90,000 to 120,000.

Chromium forms chrome steel, and with the further addition of nickel is called chrome nickel steel. This increases the hardness to a high degree and adds strength without much decrease in ductility. Chrome steels are used for high-speed cutting tools, armor plate, files, springs, safes, dies, etc.

Manganese has been mentioned under *Steel*. Its alloy is much used for high-speed cutting tools, the steel hardening when cooled in the air and being called self-hardening.

Molybdenum is used to increase the hardness to a high degree and makes the steel suitable for high-speed cutting and gives it self-hardening properties.

Nickel, with which is often combined chromium, increases the strength, springiness and toughness and helps to prevent corrosion.

Silicon has already been described. It suits the metal for use in high-speed tools.

Silver added to steel has many of the properties of nickel.

Tungsten increases the hardness without making the steel brittle. This makes the steel well suited for gas engine valves as it resists corrosion and pitting. Chromium and manganese are often used in combination with tungsten when high-speed cutting tools are made.

Vanadium as an alloy increases the elastic limit, making the steel stronger, tougher and harder. It also makes the steel able to stand much bending and vibration.

Copper.—The principal copper alloys include brass, bronze, german silver and gun metal.

Brass is composed of approximately one-third zinc and two-thirds copper. It is used for bearings and bushings where the speeds are slow and the loads rather heavy for the bearing size. It also finds use in washers, collars and forms of brackets where the metal should be non-magnetic, also for many highly finished parts.

Brass is about one-third as good an electrical conductor as copper, is slightly heavier than steel and has a tensile strength of 15,000 pounds when cast and about 75,000 to 100,000 pounds when drawn into wire.

Bronze is composed of copper and tin in various proportions, according to the use to which it is to be put. There will always be from six-tenths to nine-tenths of copper in the mixture. Bronze is used for bearings, bushings, thrust washers, brackets and gear wheels. It is heavier than steel, about 1/15 as good an electrical conductor as pure copper and has a tensile strength of 30,000 to 60,000 pounds.

Aluminum bronze, composed of copper, zinc and aluminum has high tensile strength combined with ductility and is used for parts requiring this combination.

Bearing bronze is a variable material, its composition and proportion depending on the maker and the use for which it is designed. It usually contains from 75 to 85 per cent of copper combined with one or more elements, such as tin, zinc, antimony and lead.

White metal is one form of bearing bronze containing over 80 per cent of zinc together with copper, tin, antimony and lead. Another form is made with nearly 90 per cent of tin combined with copper and antimony.

Gun metal bronze is made from 90 per cent copper with 10 per cent of tin and is used for heavy bearings, brackets and highly finished parts.

Phosphor bronze is used for very strong castings and bearings. It is similar to gun metal bronze, except that about 1-1/2 per cent of phosphorus has been added.

Manganese bronze contains about 1 per cent of manganese and is used for parts requiring great strength while being free from corrosion.

German silver is made from 60 per cent of copper with 20 per cent each of zinc and nickel. Its high electrical resistance makes it valuable for regulating devices and rheostats.

Tin is the principal part of *babbitt* and *solder*. A commonly used babbitt is composed of 89 per cent tin, 8 per cent antimony and 3 per cent of copper. A grade suitable for repairing is made from 80 per cent of lead and 20 per cent antimony. This last formula should not be used for particular work or heavy loads, being more suitable for spacers. Innumerable proportions of metals are marketed under the name of babbitt.

Solder is made from 50 per cent tin and 50 per cent lead, this grade being called "half-and-half." Hard solder is made from two-thirds tin and one-third lead.

Aluminum forms many different alloys, giving increased strength to whatever metal it unites with.

Aluminum brass is composed of approximately 65 per cent copper, 30 per cent zinc and 5 per cent aluminum. It forms a metal with high tensile strength while being ductile and malleable.

Aluminum zinc is suitable for castings which must be stiff and hard.

Nickel aluminum has a tensile strength of 40,000 pounds per square inch.

Magnalium is a silver-white alloy of aluminum with from 5 to 20 per cent of magnesium, forming a metal even lighter than aluminum and strong enough to be used in making high-speed gasoline engines.

HEAT TREATMENT OF STEEL

The processes of heat treatment are designed to suit the steel for various purposes by changing the size of the grain in the metal, therefore the strength; and by altering the chemical composition of the alloys in the metal to give it different physical properties. Heat treatment, as applied in ordinary shop work, includes the three processes of annealing, hardening and tempering, each designed to accomplish a certain definite result.

All of these processes require that the metal treated be gradually brought to a certain predetermined degree of heat which shall be uniform throughout the piece being handled and, from this point, cooled according to certain rules, the selection of which forms the difference in the three methods.

Annealing.—This is the process which relieves all internal strains and distortion in the metal and softens it so that it may more easily be cut, machined or bent to the required form. In some cases annealing is used only to relieve the strains, this being the case after forging or welding operations have been performed. In other cases it is only desired to soften the metal sufficiently that it may be handled easily. In some cases both of these things must be accomplished, as after a piece has been forged and must be machined. No matter what the object, the procedure is the same.

The steel to be annealed must first be heated to a dull red. This heating should be done slowly so that all parts of the piece have time to reach the same temperature at very nearly the same time. The piece may be heated in the forge, but a much better way is to heat in an oven or furnace of some type where the work is protected against air currents, either hot or cold, and is also protected against the direct action of the fire.

[Illustration: Figure 4.—A Gaspipe Annealing Oven]

Probably the simplest of all ovens for small tools is made by placing a piece of ordinary gas pipe in the fire (Figure 4), and heating until the inside of the pipe is bright red. Parts placed in this pipe, after one end has been closed, may be brought to the desired heat without danger of cooling draughts or chemical change from the action of the fire. More elaborate ovens may be bought which use gas, fuel oils or coal to produce the heat and in which the work may be placed on trays so that the fire will not strike directly on the steel being treated.

If the work is not very important, it may be withdrawn from the fire or oven, after heating to the desired point, and allowed to cool in the air until all traces of red have disappeared when held in a dark place. The work should be held where it is reasonably free from cold air currents. If, upon touching a pine stick to the piece being annealed, the wood does not smoke, the work may then be cooled in water.

Better annealing is secured and harder metal may be annealed if the cooling is extended over a number of hours by placing the work in a bed of non-heat-conducting material, such as ashes, charred bone, asbestos fibre, lime, sand or fire clay. It should be well covered with the heat retaining material and allowed to remain until cool. Cooling may be accomplished by allowing the fire in an oven or furnace to die down and go out, leaving the work inside the oven with all openings closed. The greater the time taken for gradual cooling from the red heat, the more perfect will be the results of the annealing.

While steel is annealed by slow cooling, copper or brass is annealed by bringing to a low red heat and quickly plunging into cold water.

Hardening.—Steel is hardened by bringing to a proper temperature, slowly and evenly as for annealing, and then cooling more or less quickly, according to the grade of steel being handled. The degree of hardening is determined by the kind of steel, the temperature from which the metal is cooled and the temperature and nature of the bath into which it is plunged for cooling.

Steel to be hardened is often heated in the fire until at some heat around 600 to 700 degrees is reached, then placed in a heating bath of molten lead, heated mercury, fused cyanate of potassium, etc., the heating bath itself being kept at the proper temperature by fires acting on it. While these baths have the advantage of heating the metal evenly and to exactly the temperature desired throughout without any part becoming over or under heated, their disadvantages consist of the fact that their materials and the fumes are poisonous in most all cases, and if not poisonous, are extremely disagreeable.

The degree of heat that a piece of steel must be brought to in order that it may be hardened depends on the percentage of carbon in the steel. The greater the percentage of carbon, the lower the heat necessary to harden.

[Illustration: Figure 5. — Cooling the Test Bar for Hardening]

To find the proper heat from which any steel must be cooled, a simple test may be carried out provided a sample of the steel, about six inches long can be secured. One end of this test bar should be heated almost to its melting point, and held at this heat until the other end just turns red. Now cool the piece in water by plunging it so that both ends enter at the same time (Figure 5), that is, hold it parallel with the surface of the water when plunged in. This serves the purpose of cooling each point along the bar from a different heat. When it has cooled in the water remove the piece and break it at short intervals, about 1/2 inch, along its length. The point along the test bar which was cooled from the best possible temperature will show a very fine smooth grain and the piece cannot be cut by a file at this point. It will be necessary to remember the exact color of that point when taken from the fire, making another test if necessary, and heat all pieces of this same steel to this heat. It will be necessary to have the cooling bath always at the same temperature, or the results cannot be alike.

While steel to be hardened is usually cooled in water, many other liquids may be used. If cooled in strong brine, the heat will be extracted much quicker, and the degree of hardness will be greater. A still greater degree of hardness is secured by cooling in a bath of mercury. Care should be used with the mercury bath, as the fumes that arise are poisonous.

Should toughness be desired, without extreme hardness, the steel may be cooled in a bath of lard oil, neatsfoot oil or fish oil. To secure a result between water and oil, it is customary to place a thick layer of oil on top of water. In cooling, the piece will pass through the oil first, thus avoiding the sudden shock of the cold water, yet producing a degree of hardness almost as great as if the oil were not used.

It will, of course, be necessary to make a separate test for each cooling medium used. If the fracture of the test piece shows a coarse grain, the steel was too hot at that point; if the fracture can be cut with a file, the metal was not hot enough at that point.

When hardening carbon tool steel its heat should be brought to a cherry red, the exact degree of heat depending on the amount of carbon and the test made, then plunged into water and held there until all hissing sound and vibration ceases. Brine may be used for this purpose; it is even better than plain water. As soon as the hissing stops, remove the work from the water or brine and plunge in oil for complete cooling.

[Illustration: Figure 6.—Cooling the Tool for Tempering]

In hardening high-speed tool steel, or air hardening steels, the tool should be handled as for carbon steel, except that after the body reaches a cherry red, the cutting point must be quickly brought to a white heat, almost melting, so that it seems ready for welding. Then cool in an oil bath or in a current of cool air.

Hardening of copper, brass and bronze is accomplished by hammering or working them while cold.

Tempering is the process of making steel tough after it has been hardened, so that it will hold a cutting edge and resist cracking. Tempering makes the grain finer and the metal stronger. It does not affect the hardness, but increases the elastic limit and reduces the brittleness of the steel. In that tempering is usually performed immediately after hardening, it might be considered as a continuation of the former process.

The work or tool to be tempered is slowly heated to a cherry red and the cutting end is then dipped into water to a depth of 1/2 to 3/4 inch above the point (Figure 6). As soon as the point cools, still leaving the tool red above the part in water, remove the work from the bath and quickly rub the end with a fine emery cloth.

As the heat from the uncooled part gradually heats the point again, the color of the polished portion changes rapidly. When a certain color is reached, the tool should be completely immersed in the water until cold.

For lathe, planer, shaper and slotter tools, this color should be a light straw.

Reamers and taps should be cooled from an ordinary straw color.

Drills, punches and wood working tools should have a brown color.

Blue or light purple is right for cold chisels and screwdrivers.

Dark blue should be reached for springs and wood saws.

Darker colors than this, ranging through green and gray, denote that the piece has reached its ordinary temper, that is, it is partially annealed.

After properly hardening a spring by dipping in lard or fish oil, it should be held over a fire while still wet with the oil. The oil takes fire and burns off, properly tempering the spring.

Remember that self-hardening steels must never be dipped in water, and always remember for all work requiring degrees of heat, that the more carbon, the less heat.

Case Hardening.—This is a process for adding more carbon to the surface of a piece of steel, so that it will have good wear-resisting qualities, while being tough and strong on the inside. It has the effect of forming a very hard and durable skin on the surface of soft steel, leaving the inside unaffected.

The simplest way, although not the most efficient, is to heat the piece to be case hardened to a red heat and then sprinkle or rub the part of the surface to be hardened with potassium ferrocyanide. This material is a deadly poison and should be handled with care. Allow the cyanide to fuse on the surface of the metal and then plunge into water, brine or mercury. Repeating the process makes the surface harder and the hard skin deeper each time.

Another method consists of placing the piece to be hardened in a bed of powdered bone (bone which has been burned and then powdered) and cover with more powdered bone, holding the whole in an iron tray. Now heat the tray and bone with the work in an oven to a bright red heat for 30 minutes to an hour and then plunge the work into water or brine.

CHAPTER II

OXY-ACETYLENE WELDING AND CUTTING MATERIALS

Welding.—Oxy-acetylene welding is an autogenous welding process, in which two parts of the same or different metals are joined by causing the edges to melt and unite while molten without the aid of hammering or compression. When cool, the parts form one piece of metal.

The oxy-acetylene flame is made by mixing oxygen and acetylene gases in a special welding torch or blowpipe, producing, when burned, a heat of 6,300 degrees, which is more than twice the melting temperature of the common metals. This flame, while being of intense heat, is of very small size.

Cutting.—The process of cutting metals with the flame produced from oxygen and acetylene depends on the fact that a jet of oxygen directed upon hot metal causes the metal itself to burn away with great rapidity, resulting in a narrow slot through the section cut. The action is so fast that metal is not injured on either side of the cut.

Carbon Removal.—This process depends on the fact that carbon will burn and almost completely vanish if the action is assisted with a supply of pure oxygen gas. After the combustion is started with any convenient flame, it continues as long as carbon remains in the path of the jet of oxygen.

Materials.—For the performance of the above operations we require the two gases, oxygen and acetylene, to produce the flames; rods of metal which may be added to the joints while molten in order to give the weld sufficient strength and proper form, and various chemical powders, called fluxes, which assist in the flow of metal and in doing away with many of the impurities and other objectionable features.

Instruments. — To control the combustion of the gases and add to the convenience of the operator a number of accessories are required.

The pressure of the gases in their usual containers is much too high for their proper use in the torch and we therefore need suitable valves which allow the gas to escape from the containers when wanted, and other specially designed valves which reduce the pressure. Hose, composed of rubber and fabric, together with suitable connections, is used to carry the gas to the torch.

The torches for welding and cutting form a class of highly developed instruments of the greatest accuracy in manufacture, and must be thoroughly understood by the welder. Tables, stands and special supports are provided for holding the work while being welded, and in order to handle the various metals and allow for their peculiarities while heated use is made of ovens and torches for preheating. The operator requires the protection of goggles, masks, gloves and appliances which prevent undue radiation of the heat.

Torch Practice. — The actual work of welding and cutting requires preliminary preparation in the form of heat treatment for the metals, including preheating, annealing and tempering. The surfaces to be joined must be properly prepared for the flame, and the operation of the torches for best results requires careful and correct regulation of the gases and the flame produced.

Finally, the different metals that are to be welded require special treatment for each one, depending on the physical and chemical characteristics of the material.

It will thus be seen that the apparently simple operations of welding and cutting require special materials, instruments and preparation on the part of the operator and it is a proved fact that failures, which have been attributed to the method, are really due to lack of these necessary qualifications.

OXYGEN

Oxygen, the gas which supports the rapid combustion of the acetylene in the torch flame, is one of the elements of the air. It is the cause and the active agent of all combustion that takes place in the atmosphere. Oxygen was first discovered as a separate gas in 1774, when it was produced by heating red oxide of mercury and was given its present name by the famous chemist, Lavoisier.

Oxygen is prepared in the laboratory by various methods, these including the heating of chloride of lime and peroxide of cobalt mixed in a retort, the heating of chlorate of potash, and the separation of water into its elements, hydrogen and oxygen, by the passage of an electric current. While the last process is used on a large scale in commercial work, the others are not practical for work other than that of an experimental or temporary nature.

This gas is a colorless, odorless, tasteless element. It is sixteen times as heavy as the gas hydrogen when measured by volume under the same temperature and pressure. Under all ordinary conditions oxygen remains in a gaseous form, although it turns to a liquid when compressed to 4,400 pounds to the square inch and at a temperature of 220° below zero.

Oxygen unites with almost every other element, this union often taking place with great heat and much light, producing flame. Steel and iron will burn rapidly when placed in this gas if the combustion is started with a flame of high heat playing on the metal. If the end of a wire is heated bright red and quickly plunged into a jar containing this gas, the wire will burn away with a dazzling light and be entirely consumed except for the molten drops that separate themselves. This property of oxygen is used in oxy-acetylene cutting of steel.

The combination of oxygen with other substances does not necessarily cause great heat, in fact the combination may be so slow and gradual that the change of temperature can not be noticed. An example of this slow combustion, or oxidation, is found in the conversion of iron into rust as the metal combines with the active gas. The respiration of human beings and animals is a form of slow combustion and is the source of animal heat. It is a general rule that the process of oxidation takes place with increasing rapidity as the temperature of the body being acted upon rises. Iron and steel at a red heat oxidize rapidly with the formation of a scale and possible damage to the metal.

Air. — Atmospheric air is a mixture of oxygen and nitrogen with traces of carbonic acid gas and water vapor. Twenty-one per cent of the air, by volume, is oxygen and the remaining seventy-nine per cent is the inactive gas, nitrogen. But for the presence of the nitrogen, which deadens the action of the other gas, combustion would take place at a destructive rate and be beyond human control in almost all cases. These two gases exist simply as a mixture to form the air and are not chemically combined. It is therefore a comparatively simple matter to separate them with the processes now available.

Water. — Water is a combination of oxygen and hydrogen, being composed of exactly two volumes of hydrogen to one volume of oxygen. If these two gases be separated from each other and then allowed to mix in these proportions they unite with explosive violence and form water. Water itself may be separated into the gases by any one of several means, one making use of a temperature of 2,200° to bring about this separation.

[Illustration: Figure 7. — Obtaining Oxygen by Electrolysis]

The easiest way to separate water into its two parts is by the process called electrolysis (Figure 7). Water, with which has been mixed a small quantity of acid, is placed in a vat through the walls of which enter the platinum tipped ends of two electrical conductors, one positive and the other negative.

Tubes are placed directly above these wire terminals in the vat, one tube being over each electrode and separated from each other by some distance. With the passage of an electric current from one wire terminal to the other, bubbles of gas rise from each and pass into the tubes. The gas that comes from the negative terminal is hydrogen and that from the positive pole is oxygen, both gases being almost pure if the work is properly conducted. This method produces electrolytic oxygen and electrolytic hydrogen.

The Liquid Air Process. — While several of the foregoing methods of securing oxygen are successful as far as this result is concerned, they are not profitable from a financial standpoint. A process for separating oxygen from the nitrogen in the air has been brought to a high state of perfection and is now supplying a major part of this gas for oxy-acetylene welding. It is known as the Linde process and the gas is distributed by the Linde Air Products Company from its plants and warehouses located in the large cities of the country.

The air is first liquefied by compression, after which the gases are separated and the oxygen collected. The air is purified and then compressed by successive stages in powerful machines designed for this purpose until it reaches a pressure of about 3,000 pounds to the square inch. The large amount of heat produced is absorbed by special coolers during the process of compression. The highly compressed air is then dried and the temperature further reduced by other coolers.

The next point in the separation is that at which the air is introduced into an apparatus called an interchanger and is allowed to escape through a valve, causing it to turn to a liquid. This liquid air is sprayed onto plates and as it falls, the nitrogen return to its gaseous state and leaves the oxygen to run to the bottom of the container. This liquid oxygen is then allowed to return to a gas and is stored in large gasometers or tanks.

The oxygen gas is taken from the storage tanks and compressed to approximately 1,800 pounds to the square inch, under which pressure it is passed into steel cylinders and made ready for delivery to the customer. This oxygen is guaranteed to be ninety-seven per cent pure.

Another process, known as the Hildebrandt process, is coming into use in this country. It is a later process and is used in Germany to a much greater extent than the Linde process. The Superior Oxygen Co. has secured the American rights and has established several plants.

Oxygen Cylinders. — Two sizes of cylinders are in use, one containing 100 cubic feet of gas when it is at atmospheric pressure and the other containing 250 cubic feet under similar conditions. The cylinders are made from one piece of steel and are without seams. These containers are tested at double the pressure of the gas contained to insure safety while handling.

One hundred cubic feet of oxygen weighs nearly nine pounds (8.921), and therefore the cylinders will weigh practically nine pounds more when full than after emptying, if of the 100 cubic feet size. The large cylinders weigh about eighteen and one-quarter pounds more when full than when empty, making approximately 212 pounds empty and 230 pounds full.

The following table gives the number of cubic feet of oxygen remaining in the cylinders according to various gauge pressures from an initial pressure of 1,800 pounds. The amounts given are not exactly correct as this would necessitate lengthy calculations which would not make great enough difference to affect the practical usefulness of the table:

Cylinder of 100 Cu. Ft. Capacity at 68° Fahr.

Gauge Pressure	Volume Remaining	Gauge Pressure	Volume Remaining
1800	100	700	39
1620	90	500	28
1440	80	300	17
1260	70	100	6
1080	60	18	1
900	50	9	1/2

Cylinder of 250 Cu. Ft. Capacity at 68° Fahr.

Gauge Pressure	Volume Remaining	Gauge Pressure	Volume Remaining
1800	250	700	97
1620	225	500	70
1440	200	300	42
1260	175	100	15
1080	150	18	8
900	125	9	1-1/4

The temperature of the cylinder affects the pressure in a large degree, the pressure increasing with a rise in temperature and falling with a fall in temperature. The variation for a 100 cubic foot cylinder at various temperatures is given in the following tabulation:

At 150° Fahr.......................... 2090 pounds.
At 100° Fahr.......................... 1912 pounds.
At 80° Fahr.......................... 1844 pounds.
At 68° Fahr.......................... 1800 pounds.
At 50° Fahr.......................... 1736 pounds.
At 32° Fahr.......................... 1672 pounds.
At 0 Fahr.......................... 1558 pounds.
At -10° Fahr.......................... 1522 pounds.

Chlorate of Potash Method. — In spite of its higher cost and the inferior gas produced, the chlorate of potash method of producing oxygen is used to a limited extent when it is impossible to secure the gas in cylinders.

[Illustration: Figure 8. — Oxygen from Chlorate of Potash]

An iron retort (Figure 8) is arranged to receive about fifteen pounds of chlorate of potash mixed with three pounds of manganese dioxide, after which the cylinder is closed with a tight cap, clamped on. This retort is carried above a burner using fuel gas or other means of generating heat and this burner is lighted after the chemical charge is mixed and compressed in the tube.

The generation of gas commences and the oxygen is led through water baths which wash and cool it before storing in a tank connected with the plant. From this tank the gas is compressed into portable cylinders at a pressure of about 300 pounds to the square inch for use as required in welding operations.

Each pound of chlorate of potash liberates about three cubic feet of oxygen, and taking everything into consideration, the cost of gas produced in this way is several times that of the purer product secured by the liquid air process.

These chemical generators are oftentimes a source of great danger, especially when used with or near the acetylene gas generator, as is sometimes the case with cheap portable outfits. Their use should not be tolerated when any other method is available, as the danger from accident alone should prohibit the practice except when properly installed and cared for away from other sources of combustible gases.

ACETYLENE

In 1862 a chemist, Woehler, announced the discovery of the preparation of acetylene gas from calcium carbide, which he had made by heating to a high temperature a mixture of charcoal with an alloy of zinc and calcium. His product would decompose water and yield the gas. For nearly thirty years these substances were neglected, with the result that acetylene was practically unknown, and up to

1892 an acetylene flame was seen by very few persons and its possibilities were not dreamed of. With the development of the modern electric furnace the possibility of calcium carbide as a commercial product became known.

In the above year, Thomas L. Willson, an electrical engineer of Spray, North Carolina, was experimenting in an attempt to prepare metallic calcium, for which purpose he employed an electric furnace operating on a mixture of lime and coal tar with about ninety-five horse power. The result was a molten mass which became hard and brittle when cool. This apparently useless product was discarded and thrown in a nearby stream, when, to the astonishment of onlookers, a large volume of gas was immediately liberated, which, when ignited, burned with a bright and smoky flame and gave off quantities of soot. The solid material proved to be calcium carbide and the gas acetylene.

Thus, through the incidental study of a by-product, and as the result of an accident, the possibilities in carbide were made known, and in the spring of 1895 the first factory in the world for the production of this substance was established by the Willson Aluminum Company.

When water and calcium carbide are brought together an action takes place which results in the formation of acetylene gas and slaked lime.

CARBIDE

Calcium carbide is a chemical combination of the elements carbon and calcium, being dark brown, black or gray with sometimes a blue or red tinge. It looks like stone and will only burn when heated with oxygen.

Calcium carbide may be preserved for any length of time if protected from the air, but the ordinary moisture in the atmosphere gradually affects it until nothing remains but slaked lime. It always possesses a penetrating odor, which is not due to the carbide itself but to the fact that it is being constantly affected by moisture and producing small quantities of acetylene gas.

This material is not readily dissolved by liquids, but if allowed to come in contact with water, a decomposition takes place with the evolution of large quantities of gas. Carbide is not affected by shock, jarring or age.

A pound of absolutely pure carbide will yield five and one-half cubic feet of acetylene. Absolute purity cannot be attained commercially, and in practice good carbide will produce from four and one-half to five cubic feet for each pound used.

Carbide is prepared by fusing lime and carbon in the electric furnace under a heat in excess of 6,000 degrees Fahrenheit. These materials are among the most difficult to melt that are known. Lime is so infusible that it is frequently employed for the materials of crucibles in which the highest melting metals are fused, and for the pencils in the calcium light because it will stand extremely high temperatures.

Carbon is the material employed in the manufacture of arc light electrodes and other electrical appliances that must stand extreme heat. Yet these two substances are forced into combination in the manufacture of calcium carbide. It is the excessively high temperature attainable in the electric furnace that causes this combination and not any effect of the electricity other than the heat produced.

A mixture of ground coke and lime is introduced into the furnace through which an electric arc has been drawn. The materials unite and form an ingot of very pure carbide surrounded by a crust of less purity. The poorer crust is rejected in breaking up the mass into lumps which are graded according to their size. The largest size is 2 by 3-1/2 inches and is called "lump," a medium size is 1/2 by 2 inches and is called "egg," an intermediate size for certain types of generators is 3/8 by 1-1/4 inches and called "nut," and the finely crushed pieces for use in still other types of generators are 1/12 by 1/4 inch in size and are called "quarter." Instructions as to the size best suited to different generators are furnished by the makers of those instruments.

These sizes are packed in air-tight sheet steel drums containing 100 pounds each. The Union Carbide Company of Chicago and New York, operating under patents, manufactures and distributes the supply of calcium carbide for the entire United States. Plants for this manufacture are established at Niagara Falls, New York, and

Sault Ste. Marie, Michigan. This company maintains a system of warehouses in more than one hundred and ten cities, where large stocks of all sizes are carried.

The National Board of Fire Underwriters gives the following rules for the storage of carbide:

Calcium carbide in quantities not to exceed six hundred pounds may be stored, when contained in approved metal packages not to exceed one hundred pounds each, inside insured property, provided that the place of storage be dry, waterproof and well ventilated and also provided that all but one of the packages in any one building shall be sealed and that seals shall not be broken so long as there is carbide in excess of one pound in any other unsealed package in the building.

Calcium carbide in quantities in excess of six hundred pounds must be stored above ground in detached buildings, used exclusively for the storage of calcium carbide, in approved metal packages, and such buildings shall be constructed to be dry, waterproof and well ventilated.

Properties of Acetylene. — This gas is composed of twenty-four parts of carbon and two parts of hydrogen by weight and is classed with natural gas, petroleum, etc., as one of the hydrocarbons. This gas contains the highest percentage of carbon known to exist in any combination of this form and it may therefore be considered as gaseous carbon. Carbon is the fuel that is used in all forms of combustion and is present in all fuels from whatever source or in whatever form. Acetylene is therefore the most powerful of all fuel gases and is able to give to the torch flame in welding the highest temperature of any flame.

Acetylene is a colorless and tasteless gas, possessed of a peculiar and penetrating odor. The least trace in the air of a room is easily noticed, and if this odor is detected about an apparatus in operation, it is certain to indicate a leakage of gas through faulty piping, open valves, broken hose or otherwise. This leakage must be prevented before proceeding with the work to be done.

All gases which burn in air will, when mixed with air previous to ignition, produce more or less violent explosions, if fired. To this rule acetylene is no exception. One measure of acetylene and twelve and one-half of air are required for complete combustion; this is therefore the proportion for the most perfect explosion. This is not the only possible mixture that will explode, for all proportions from three to thirty per cent of acetylene in air will explode with more or less force if ignited.

The igniting point of acetylene is lower than that of coal gas, being about 900 degrees Fahrenheit as against eleven hundred degrees for coal gas. The gas issuing from a torch will ignite if allowed to play on the tip of a lighted cigar.

It is still further true that acetylene, at some pressures, greater than normal, has under most favorable conditions for the effect, been found to explode; yet it may be stated with perfect confidence that under no circumstances has anyone ever secured an explosion in it when subjected to pressures not exceeding fifteen pounds to the square inch.

Although not exploded by the application of high heat, acetylene is injured by such treatment. It is partly converted, by high heat, into other compounds, thus lessening the actual quantity of the gas, wasting it and polluting the rest by the introduction of substances which do not belong there. These compounds remain in part with the gas, causing it to burn with a persistent smoky flame and with the deposit of objectionable tarry substances. Where the gas is generated without undue rise of temperature these difficulties are avoided.

Purification of Acetylene. — Impurities in this gas are caused by impurities in the calcium carbide from which it is made or by improper methods and lack of care in generation. Impurities from the material will be considered first.

Impurities in the carbide may be further divided into two classes: those which exert no action on water and those which act with the water to throw off other gaseous products which remain in the acetylene. Those impurities which exert no action on the water consist of coke that has not been changed in the furnace and sand and some other substances which are harmless except that they increase the ash left after the acetylene has been generated.

An analysis of the gas coming from a typical generator is as follows:

	Per cent
Acetylene	99.36
Oxygen	.08
Nitrogen	.11
Hydrogen	.06
Sulphuretted Hydrogen	.17
Phosphoretted Hydrogen	.04
Ammonia	.10
Silicon Hydride	.03
Carbon Monoxide	.01
Methane	.04

The oxygen, nitrogen, hydrogen, methane and carbon monoxide are either harmless or are present in such small quantities as to be neglected. The phosphoretted hydrogen and silicon hydride are self-inflammable gases when exposed to the air, but their quantity is so very small that this possibility may be dismissed. The ammonia and sulphuretted hydrogen are almost entirely dissolved by the water used in the gas generator. The surest way to avoid impure gas is to use high-grade calcium carbide in the generator and the carbide of American manufacture is now so pure that it never causes trouble.

The first and most important purification to which the gas is subjected is its passage through the body of water in the generator as it bubbles to the top. It is then filtered through felt to remove the solid particles of lime dust and other impurities which float in the gas.

Further purification to remove the remaining ammonia, sulphuretted hydrogen and phosphorus containing compounds is accomplished by chemical means. If this is considered necessary it can be easily accomplished by readily available purifying apparatus which can be attached to any generator or inserted between the generator and torch outlets. The following mixtures have been used.

"*Heratol,*" a solution of chromic acid or sulphuric acid absorbed in porous earth.

"*Acagine*," a mixture of bleaching powder with fifteen per cent of lead chromate.

"*Puratylene*," a mixture of bleaching powder and hydroxide of lime, made very porous, and containing from eighteen to twenty per cent of active chlorine.

"*Frankoline*," a mixture of cuprous and ferric chlorides dissolved in strong hydrochloric acid absorbed in infusorial earth.

A test for impure acetylene gas is made by placing a drop of ten per cent solution of silver nitrate on a white blotter and holding the paper in a stream of gas coming from the torch tip. Blackening of the paper in a short length of time indicates impurities.

Acetylene in Tanks.—Acetylene is soluble in water to a very limited extent, too limited to be of practical use. There is only one liquid that possesses sufficient power of containing acetylene in solution to be of commercial value, this being the liquid acetone. Acetone is produced in various ways, oftentimes from the distillation of wood. It is a transparent, colorless liquid that flows with ease. It boils at 133° Fahrenheit, is inflammable and burns with a luminous flame. It has a peculiar but rather agreeable odor.

Acetone dissolves twenty-four times its own bulk of acetylene at ordinary atmospheric pressure. If this pressure is increased to two atmospheres, 14.7 pounds above ordinary pressure, it will dissolve just twice as much of the gas and for each atmosphere that the pressure is increased it will dissolve as much more.

If acetylene be compressed above fifteen pounds per square inch at ordinary temperature without first being dissolved in acetone a danger is present of self-ignition. This danger, while practically nothing at fifteen pounds, increases with the pressure until at forty atmospheres it is very explosive. Mixed with acetone, the gas loses this dangerous property and is safe for handling and transportation. As acetylene is dissolved in the liquid the acetone increases its volume slightly so that when the gas has been drawn out of a closed tank a space is left full of free acetylene.

This last difficulty is removed by first filling the cylinder or tank with some porous material, such as asbestos, wood charcoal, infusorial earth, etc. Asbestos is used in practice and by a system of packing and supporting the absorbent material no space is left for the free gas, even when the acetylene has been completely withdrawn.

The acetylene is generated in the usual way and is washed, purified and dried. Great care is used to make the gas as free as possible from all impurities and from air. The gas is forced into containers filled with acetone as described and is compressed to one hundred and fifty pounds to the square inch. From these tanks it is transferred to the smaller portable cylinders for consumers' use.

The exact volume of gas remaining in a cylinder at atmospheric temperature may be calculated if the weight of the cylinder empty is known. One pound of the gas occupies 13.6 cubic feet, so that if the difference in weight between the empty cylinder and the one considered be multiplied by 13.6. the result will be the number of cubic feet of gas contained.

The cylinders contain from 100 to 500 cubic feet of acetylene under pressure. They cannot be filled with the ordinary type of generator as they require special purifying and compressing apparatus, which should never be installed in any building where other work is being carried on, or near other buildings which are occupied, because of the danger of explosion.

Dissolved acetylene is manufactured by the Prest-O-Lite Company, the Commercial Acetylene Company and the Searchlight Gas Company and is distributed from warehouses in various cities.

These tanks should not be discharged at a rate per hour greater than one-seventh of their total capacity, that is, from a tank of 100 cubic feet capacity, the discharge should not be more than fourteen cubic feet per hour. If discharge is carried on at an excessive rate the acetone is drawn out with the gas and reduces the heat of the welding flame.

For this reason welding should not be attempted with cylinders designed for automobile and boat lighting. When the work demands a greater delivery than one of the larger tanks will give, two or more tanks may be connected with a special coupler such as may be secured from the makers and distributers of the gas. These couplers may be arranged for two, three, four or five tanks in one battery by removing the plugs on the body of the coupler and attaching additional connecting pipes. The coupler body carries a pressure gauge and the valve for controlling the pressure of the gas as it flows to the welding torches. The following capacities should be provided for:

Acetylene Consumption Combined Capacity of of Torches per Hour Cylinders in Use Up to 15 feet……………………100 cubic feet 16 to 30 feet………………….200 cubic feet 31 to 45 feet…………………..300 cubic feet 46 to 60 feet…………………..400 cubic feet 61 to 75 feet…………………..500 cubic feet

WELDING RODS

The best welding cannot be done without using the best grade of materials, and the added cost of these materials over less desirable forms is so slight when compared to the quality of work performed and the waste of gases with inferior supplies, that it is very unprofitable to take any chances in this respect. The makers of welding equipment carry an assortment of supplies that have been standardized and that may be relied upon to produce the desired result when properly used. The safest plan is to secure this class of material from the makers.

Welding rods, or welding sticks, are used to supply the additional metal required in the body of the weld to replace that broken or cut away and also to add to the joint whenever possible so that the work may have the same or greater strength than that found in the original piece. A rod of the same material as that being welded is used when both parts of the work are the same. When dissimilar metals are to be joined rods of a composition suited to the work are employed.

These filling rods are required in all work except steel of less than 16 gauge. Alloy iron rods are used for cast iron. These rods have a high silicon content, the silicon reacting with the carbon in the iron to produce a softer and more easily machined weld than would otherwise be the case. These rods are often made so that they melt at a slightly lower point than cast iron. This is done for the reason that when the part being welded has been brought to the fusing heat by the torch, the filling material can be instantly melted in without allowing the parts to cool. The metal can be added faster and more easily controlled.

Rods or wires of Norway iron are used for steel welding in almost all cases. The purity of this grade of iron gives a homogeneous, soft weld of even texture, great ductility and exceptionally good machining qualities. For welding heavy steel castings, a rod of rolled carbon steel is employed. For working on high carbon steel, a rod of the steel being welded must be employed and for alloy steels, such as nickel, manganese, vanadium, etc., special rods of suitable alloy composition are preferable.

Aluminum welding rods are made from this metal alloyed to give the even flowing that is essential. Aluminum is one of the most difficult of all the metals to handle in this work and the selection of the proper rod is of great importance.

Brass is filled with brass wire when in small castings and sheets. For general work with brass castings, manganese bronze or Tobin bronze may be used.

Bronze is welded with manganese bronze or Tobin bronze, while copper is filled with copper wire.

These welding rods should always be used to fill the weld when the thickness of material makes their employment necessary, and additional metal should always be added at the weld when possible as the joint cannot have the same strength as the original piece if made or dressed off flush with the surfaces around the weld. This is true because the metal welded into the joint is a casting and will never have more strength than a casting of the material used for filling.

Great care should be exercised when adding metal from welding rods to make sure that no metal is added at a point that is not itself melted and molten when the addition is made. When molten metal is placed upon cooler surfaces the result is not a weld but merely a sticking together of the two parts without any strength in the joint.

FLUXES

Difficulty would be experienced in welding with only the metal and rod to work with because of the scale that forms on many materials under heat, the oxides of other metals and the impurities found in almost all metals. These things tend to prevent a perfect joining of the metals and some means are necessary to prevent their action.

Various chemicals, usually in powder form, are used to accomplish the result of cleaning the weld and making the work of the operator less difficult. They are called fluxes.

A flux is used to float off physical impurities from the molten metal; to furnish a protecting coating around the weld; to assist in the removal of any objectionable oxide of the metals being handled; to lower the temperature at which the materials flow; to make a cleaner weld and to produce a better quality of metal in the finished work.

The flux must be of such composition that it will accomplish the desired result without introducing new difficulties. They may be prepared by the operator in many cases or may be secured from the makers of welding apparatus, the same remarks applying to their quality as were made regarding the welding rods, that is, only the best should be considered.

The flux used for cast iron should have a softening effect and should prevent burning of the metal. In many cases it is possible and even preferable to weld cast iron without the use of a flux, and in any event the smaller the quantity used the better the result should be. Flux should not be added just before the completion of the work because the heat will not have time to drive the added elements out of the metal or to incorporate them with the metal properly.

Aluminum should never be welded without using a flux because of the oxide formed. This oxide, called alumina, does not melt until a heat of 5,000° Fahrenheit is reached, four times the heat needed to melt the aluminum itself. It is necessary that this oxide be broken down or dissolved so that the aluminum may have a chance to flow together. Copper is another metal that requires a flux because of its rapid oxidation under heat.

While the flux is often thrown or sprinkled along the break while welding, much better results will be obtained by dipping the hot end of the welding rod into the flux whenever the work needs it. Sufficient powder will stick on the end of the rod for all purposes, and with some fluxes too much will adhere. Care should always be used to avoid the application of excessive flux, as this is usually worse than using too little.

SUPPLIES AND FIXTURES

Goggles.—The oxy-acetylene torch should not be used without the protection to the eyes afforded by goggles. These not only relieve unnecessary strain, but make it much easier to watch the exact progress of the work with the molten metal. The difficulty of protecting the sight while welding is even greater than when cutting metal with the torch.

Acetylene gives a light which is nearest to sunlight of any artificial illuminant. But for the fact that this gas light gives a little more green and less blue in its composition, it would be the same in quality and practically the same in intensity. This light from the gas is almost absent during welding, being lost with the addition of the extra oxygen needed to produce the welding heat. The light that is dangerous comes from the molten metal which flows under the torch at a bright white heat.

Goggles for protection against this light and the heat that goes with it may be secured in various tints, the darker glass being for welding and the lighter for cutting. Those having frames in which the metal parts do not touch the flesh directly are most desirable because of the high temperature reached by these parts.

Gloves.—While not as necessary as are the goggles, gloves are a convenience in many cases. Those in which leather touches the hands directly are really of little value as the heat that protection is desired against makes the leather so hot that nothing is gained in comfort. Gloves are made with asbestos cloth, which are not open to this objection in so great a degree.

[Illustration: Figure 9.—Frame for Welding Stand]

Tables and Stands.—Tables for holding work while being welded (Figure 9) are usually made from lengths of angle steel welded together. The top should be rectangular, about two feet wide and two and one-half feet long. The legs should support the working surface at a height of thirty-two to thirty-six inches from the floor. Metal lattice work may be fastened or laid in the top framework and used to support a layer of firebrick bound together with a mixture of one-third cement and two-thirds fireclay. The piece being welded is braced and supported on this table with pieces of firebrick so that it will remain stationary during the operation.

Holders for supporting the tanks of gas may be made or purchased in forms that rest directly on the floor or that are mounted on wheels. These holders are quite useful where the floor or ground is very uneven.

Hose.—All permanent lines from tanks and generators to the torches are made with piping rigidly supported, but the short distance from the end of the pipe line to the torch itself is completed with a flexible hose so that the operator may be free in his movements while welding. An accident through which the gases mix in the hose and are ignited will burst this part of the equipment, with more or less painful results to the person handling it. For that reason it is well to use hose with great enough strength to withstand excessive pressure.

A poor grade of hose will also break down inside and clog the flow of gas, both through itself and through the parts of the torch. To avoid outside damage and cuts this hose is sometimes encased with coiled sheet metal. Hose may be secured with a bursting strength of more than 1,000 pounds to the square inch. Many operators prefer to distinguish between the oxygen and acetylene lines by their color and to allow this, red is used for the oxygen and black for acetylene.

Other Materials.—Sheet asbestos and asbestos fibre in flakes are used to cover parts of the work while preparing them for welding and during the operation itself. The flakes and small pieces that become detached from the large sheets are thrown into a bin where the completed small work is placed to allow slow and even cooling while protected by the asbestos.

Asbestos fibre and also ordinary fireclay are often used to make a backing or mould into a form that may be placed behind aluminum and some other metals that flow at a low heat and which are accordingly difficult to handle under ordinary methods. This forms a solid mould into which the metal is practically cast as melted by the torch so that the desired shape is secured without danger of the walls of metal breaking through and flowing away.

Carbon blocks and rods are made in various shapes and sizes so that they may be used to fill threaded holes and other places that it is desired to protect during welding. These may be secured in rods of various diameters up to one inch and in blocks of several different dimensions.

CHAPTER III

ACETYLENE GENERATORS

Acetylene generators used for producing the gas from the action of water on calcium carbide are divided into three principal classes according to the pressure under which they operate.

Low pressure generators are designed to operate at one pound or less per square inch. Medium pressure systems deliver the gas at not to exceed fifteen pounds to the square inch while high pressure types furnish gas above fifteen pounds per square inch. High pressure systems are almost unknown in this country, the medium pressure type being often referred to as "high pressure."

Another important distinction is formed by the method of bringing the carbide and water together. The majority of those now in use operate by dropping small quantities of carbide into a large volume of water, allowing the generated gas to bubble up through the water before being collected above the surface. This type is known as the "carbide to water" generator.

A less used type brings a measured and small quantity of water to a comparatively large body of the carbide, the gas being formed and collected from the chamber in which the action takes place. This is called the "water to carbide" type. Another way of expressing the difference in feed is that of designating the two types as "carbide feed" for the former and "water feed" for the latter.

A further division of the carbide to water machines is made by mentioning the exact method of feeding the carbide. One type, called "gravity feed" operates by allowing the carbide to escape and fall by the action of its own weight, or gravity; the other type, called "forced feed," includes a separate mechanism driven by power. This mechanism feeds definite amounts of the carbide to the water as required by the demands on the generator. The action of either feed is controlled by the withdrawal of gas from the generator, the aim being to supply sufficient carbide to maintain a nearly constant supply.

Generator Requirements. — The qualities of a good generator are outlined as follows: [Footnote: See Pond's "Calcium Carbide and Acetylene."]

It must allow no possibility of the existence of an explosive mixture in any of its parts at any time. It is not enough to argue that a mixture, even if it exists, cannot be exploded unless kindled. It is necessary to demand that a dangerous mixture can at no time be formed, even if the machine is tampered with by an ignorant person. The perfect machine must be so constructed that it shall be impossible at any time, under any circumstances, to blow it up.

It must insure cool generation. Since this is a relative term, all machines being heated somewhat during the generation of gas, this amounts to saying that a machine must heat but little. A pound of carbide decomposed by water develops the same amount of heat under all circumstances, but that heat can be allowed to increase locally to a high point, or it can be equalized by water so that no part of the material becomes heated enough to do damage.

It must be well constructed. A good generator does not need, perhaps, to be "built like a watch," but it should be solid, substantial and of good material. It should be built for service, to last and not simply to sell; anything short of this is to be avoided as unsafe and unreliable.

It must be simple. The more complicated the machine the sooner it will get out of order. Understand your generator. Know what is inside of it and beware of an apparatus, however attractive its exterior, whose interior is filled with pipes and tubes, valves and diaphragms whose functions you do not perfectly understand.

It should be capable of being cleaned and recharged and of receiving all other necessary attention without loss of gas, both for economy's sake, and more particularly to avoid danger of fire.

It should require little attention. All machines have to be emptied and recharged periodically; but the more this process is simplified and the more quickly this can be accomplished, the better.

It should be provided with a suitable indicator to designate how low the charge is in order that the refilling may be done in good season.

It should completely use up the carbide, generating the maximum amount of gas.

Overheating. — A large amount of heat is liberated when acetylene gas is formed from the union of calcium carbide and water. Overheating during this process, that is to say, an intense local heat rather than a large amount of heat well distributed, brings about the phenomenon of polymerization, converting the gas, or part of it, into oily matters, which can do nothing but harm. This tarry mass coming through the small openings in the torches causes them to become partly closed and alters the proportions of the gases to the detriment of the welding flame. The only remedy for this trouble is to avoid its cause and secure cool generation.

Overheating can be detected by the appearance of the sludge remaining after the gas has been made. Discoloration, yellow or brown, shows that there has been trouble in this direction and the resultant effects at the torches may be looked for. The abundance of water in the carbide to water machines effects this cooling naturally and is a characteristic of well designed machines of this class. It has been found best and has practically become a fundamental rule of generation that a gallon of water must be provided for each pound of carbide placed in the generator. With this ratio and a generator large enough for the number of torches to be supplied, little trouble need be looked for with overheating.

Water to Carbide Generators. — It is, of course, much easier to obtain a measured and regular flow of water than to obtain such a flow of any solid substance, especially when the solid substance is in the form of lumps, as is carbide This fact led to the use of a great many water-feed generators for all classes of work, and this type is still in common use for the small portable machines, such, for instance, as those used on motor cars for the lamps. The water-feed machine is not, however, favored for welding plants, as is the carbide feed, in spite of the greater difficulties attending the handling of the solid material.

A water-feed generator is made up of the gas producing part and a holder for the acetylene after it is made. The carbide is held in a tray formed of a number of small compartments so that the charge in each compartment is nearly equal to that in each of the others. The water is allowed to flow into one of these compartments in a volume sufficient to produce the desired amount of gas and the carbide is completely used from this one division. The water then floods the first compartment and finally overflows into the next one, where the same process is repeated. After using the carbide in this division, it is flooded in turn and the water passing on to those next in order, uses the entire charge of the whole tray.

These generators are charged with the larger sizes of carbide and are easily taken care of. The residue is removed in the tray and emptied, making the generator ready for a fresh supply of carbide.

Carbide to Water Generators. — This type also is made up of two principal parts, the generating chamber and a gas holder, the holder being part of the generating chamber or a separate device. The generator (Figure 10) contains a hopper to receive the charge of carbide and is fitted with the feeding mechanism to drop the proper amount of carbide into the water as required by the demands of the torches. The charge of carbide is of one of the smaller sizes, usually "nut" or "quarter."

Feed Mechanisms. — The device for dropping the carbide into the water is the only part of the machine that is at all complicated. This complication is brought about by the necessity of controlling the mass of carbide so that it can never be discharged into the water at an excessive rate, feeding it at a regular rate and in definite amounts, feeding it positively whenever required and shutting off the feed just as positively when the supply of gas in the holder is enough for the immediate needs.

[Illustration: Figure 10. — Carbide to Water Generator. A. Feed motor weight;
B. Carbide feed motor; C. Carbide hopper; D. Water for gas generation;
E. Agitator for loosening residuum; F. Water seal in gas bell; G. Filter;

H. Hydraulic Valve; J. Motor control levers.]

The charge of carbide is unavoidably acted upon by the water vapor in the generator and will in time become more or less pasty and sticky. This is more noticeable if the generator stands idle for a considerable length of time This condition imposes another duty on the feeding mechanism; that is, the necessity of self-cleaning so that the carbide, no matter in what condition, cannot prevent the positive action of this part of the device, especially so that it cannot prevent the supply from being stopped at the proper time.

The gas holder is usually made in the bell form so that the upper portion rises and falls with the addition to or withdrawal from the supply of gas in the holder. The rise and fall of this bell is often used to control the feed mechanism because this movement indicates positively whether enough gas has been made or that more is required. As the bell lowers it sets the feed mechanism in motion, and when the gas passing into the holder has raised the bell a sufficient distance, the movement causes the feed mechanism to stop the fall of carbide into the water. In practice, the movement of this part of the holder is held within very narrow limits.

Gas Holders. — No matter how close the adjustment of the feeding device, there will always be a slight amount of gas made after the fall of carbide is stopped, this being caused by the evolution of gas from the carbide with which water is already in contact. This action is called "after generation" and the gas holder in any type of generator must provide sufficient capacity to accommodate this excess gas. As a general rule the water to carbide generator requires a larger gas holder than the carbide to water type because of the greater amount of carbide being acted upon by the water at any one time, also because the surface of carbide presented to the moist air within the generating chamber is greater with this type.

Freezing. — Because of the rather large body of water contained in any type of generator, there is always danger of its freezing and rendering the device inoperative unless placed in a temperature above the freezing point of the water. It is, of course, dangerous and against the insurance rules to place a generator in the same room with a fire of any kind, but the room may be heated by steam or hot water coils from a furnace in another building or in another part of the same building.

When the generator is housed in a separate structure the walls should be made of materials or construction that prevents the passage of heat or cold through them to any great extent. This may be accomplished by the use of hollow tile or concrete blocks or by any other form of double wall providing air spaces between the outer and inner facings. The space between the parts of the wall may be filled with materials that further retard the loss of heat if this is necessary under the conditions prevailing.

Residue From Generators. — The sludge remaining in the carbide to water generator may be drawn off into the sewer if the piping is run at a slant great enough to give a fall that carries the whole quantity, both water and ash, away without allowing settling and consequent clogging. Generators are provided with agitators which are operated to stir the ash up with the water so that the whole mass is carried off when the drain cock is opened.

If sewer connections cannot be made in such a way that the ash is entirely carried away, it is best to run the liquid mass into a settling basin outside of the building. This should be in the form of a shallow pit which will allow the water to pass off by soaking into the ground and by evaporation, leaving the comparatively dry ash in the pit. This ash which remains is essentially slaked lime and can often be disposed of to more or less advantage to be used in mortar, whitewash, marking paths and any other use for which slaked lime is suited. The disposition of the ash depends entirely on local conditions. An average analysis of this ash is as follows:

Sand...................... 1.10 per cent. Carbon...................... 2.72 "
Oxide of iron and alumina.. 2.77 " Lime...................... 64.06 "
Water and carbonic acid.... 29.35 " — — — 100.00

GENERATOR CONSTRUCTION

The water for generating purposes is carried in the large tank-like compartment directly below the carbide chamber. See Figure 11. This water compartment is filled through a pipe of such a height that the water level cannot be brought above the proper point or else the water compartment is provided with a drain connection which accomplishes this same result by allowing an excess to flow away.

The quantity of water depends on the capacity of the generator inasmuch as there must be one gallon for each pound of carbide required. The generator should be of sufficient capacity to furnish gas under working conditions from one charge of carbide to all torches installed for at least five hours continuous use.

After calculating the withdrawal of the whole number of torches according to the work they are to do for this period of five hours the proper generator capacity may be found on the basis of one cubic foot of gas per hour for each pound of carbide. Thus if the torches were to use sixty cubic feet of gas per hour, five hours would call for three hundred cubic feet and a three hundred pound generator should be installed. Generators are rated according to their carbide capacity in pounds.

Charging.—The carbide capacity of the generator should be great enough to furnish a continuous supply of gas for the maximum operating time, basing the quantity of gas generated on four and one-half cubic feet from each pound of lump carbide and on four cubic feet from each pound of quarter, intermediate sizes being in proportion.

Generators are built in such a way that it is impossible for the acetylene to escape from the gas holding compartment during the recharging process. This is accomplished (1) by connecting the water inlet pipe opening with a shut off valve in such a way that the inlet cannot be uncovered or opened without first closing the shut off valve with the same movement of the operator; (2) by incorporating an automatic or hydraulic one-way valve so that this valve closes and acts as a check when the gas attempts to flow from the holder back to the generating chamber, or by any other means that will positively accomplish this result.

In generators having no separate gas holding chamber but carrying the supply in the same compartment in which it is generated, the gas contained under pressure is allowed to escape through vent pipes into the outside air before recharging with carbide. As in the former case, the parts are so interlocked that it is impossible to introduce carbide or water without first allowing the escape of the gas in the generator.

It is required by the insurance rules that the entire change of carbide while in the generator be held in such a way that it may be entirely removed without difficulty in case the necessity should arise.

Generators should be cleaned and recharged at regular stated intervals. This work should be done during daylight hours only and likewise all repairs should be made at such a time that artificial light is not needed. Where it is absolutely necessary to use artificial light it should be provided only by incandescent electric lamps enclosed in gas tight globes.

In charging generating chambers the old ash and all residue must first be cleaned out and the operator should be sure that no drain or other pipe has become clogged. The generator should then be filled with the required amount of water. In charging carbide feed machines be careful not to place less than a gallon of water in the water compartment for each pound of carbide to be used and the water must be brought to, but not above, the proper level as indicated by the mark or the maker's instructions. The generating chamber must be filled with the proper amount of water before any attempt is made to place the carbide in its holder. This rule must always be followed. It is also necessary that all automatic water seals and valves, as well as any other water tanks, be filled with clean water at this time.

Never recharge with carbide without first cleaning the generating chamber and completely refilling with clean water. Never test the generator or piping for leaks with any flame, and never apply flame to any open pipe or at any point other than the torch, and only to the torch after it has a welding or cutting nozzle attached. Never use a lighted match, lamp, candle, lantern, cigar or any open flame near a generator. Failure to observe these precautions is liable to endanger life and property.

Operation and Care of Generators.—The following instructions apply especially to the Davis Bournonville pressure generator, illustrated in Figure 11. The motor feed mechanism is illustrated in Figure 12.

Before filling the machine, the cover should be removed and the hopper taken out and examined to see that the feeding disc revolves freely; that no chains have been displaced or broken, and that the carbide displacer itself hangs barely free of the feeding disc when it is revolved. After replacing the cover, replace the bolts and tighten them equally, a little at a time all around the circumference of the cover—not screwing tight in one place only. Do not screw the cover down any more than is necessary to make a tight fit.

To charge the generator, proceed as follows: Open the vent valve by turning the handle which extends over the filling tube until it stands at a right angle with the generator. Open the valve in the water filling pipe, and through this fill with water until it runs out of the overflow pipe of the drainage chamber, then close the valve in the water filling pipe and vent valve. Remove the carbide filling plugs and fill the hopper with 1-1/4"x3/8" carbide ("nut" size). Then replace the plugs and the safety-locking lever chains. Now rewind the motor weight. Run the pressure up to about five pounds by raising the controlling diaphragm valve lever by hand (Figure 12, lever marked E). Then raise the blow-off lever, allowing the gas to blow off until the gauge shows about two pounds; this to clear the generator of air mixture. Then run the pressure up to about eight pounds by raising the controlling valve lever E, or until this controlling lever rests against the upper wing of the fan governor, and prevents operation of the feed motor. After this is done, the motor will operate automatically as the gas is consumed.

[Illustration: Figure 11.—Pressure Generator (Davis Bournonville). A, Feed motor weight; B, Carbide feed motor; C, Motor Control diaphragm; D, Carbide hopper; E, Carbide feed disc; F, Overflow pipe; G, Overflow pipe seal; H, Overflow pipe valve; J, Filling funnel; K, Hydraulic valve; L, Expansion chamber; M, Escape pipe; N, Feed pipe; O, Agitator for residuum; P, Residuum valve; Q, Water level]

[Illustration: Figure 12.—Feed Mechanism of Pressure Generator]

Should the pressure rise much above the blow-off point, the safety controlling diaphragm valve will operate and throw the safety clutch in interference and thus stop the motor. This interference clutch will then have to be returned to its former position before the motor will operate, but cannot be replaced before the pressure has been reduced below the blow-off point.

The parts of the feed mechanism illustrated in Figure 12 are as follows: *A*, motor drum for weight cable. *B*, carbide filling plugs. *C*, chains for connecting safety locking lever of motor to pins on the top of the carbide plugs. *D*, interference clutch of motor. *E*, lever on feed controlling diaphragm valve. *F*, lever of interference controlling diaphragm valve that operates interference clutch. *G*, feed controlling diaphragm valve. *H*, diaphragm valve controlling operation of interference clutch. *I*, interference pin to engage emergency clutch. *J*, main shaft driving carbide feeding disc. *Y*, safety locking lever.

Recharging Generator.—Turn the agitator handle rapidly for several revolutions, and then open the residuum valve, having five or six pounds gas pressure on the machine. If the carbide charge has been exhausted and the motor has stopped, there is generally enough carbide remaining in the feeding disc that can be shaken off, and fed by running the motor to obtain some pressure in the generator. The desirability of discharging the residuum with some gas pressure is because the pressure facilitates the discharge and at the same time keeps the generator full of gas, preventing air mixture to a great extent. As soon as the pressure is relieved by the withdrawal of the residuum, the vent valve should be opened, as if the pressure is maintained until all of the residuum is discharged gas would escape through the discharge valve.

Having opened the vent pipe valve and relieved the pressure, open the valve in the water filling tube. Close the residuum valve, then run in several gallons of water and revolve the agitator, after which draw out the remaining residuum; then again close the residuum valve and pour in water until it discharges from the overflow pipe of the drainage chamber. It is desirable in filling the generator to pour the water in rapidly enough to keep the filling pipe full of water, so that air will not pass in at the same time.

After the generator is cleaned and filled with water, fill with carbide and proceed in the same manner as when first charging.

Carbide Feed Mechanism.—Any form of carbide to water machine should be so designed that the carbide never falls directly from its holder into the water, but so that it must take a more or less circuitous path. This should be true, no matter what position the mechanism is in. One of the commonest types of forced feed machine carries the carbide in a hopper with slanting sides, this hopper having a large opening in the bottom through which the carbide passes to a revolving circular plate. As the pieces of carbide work out toward the edge of the plate under the influence of the mass behind them, they are thrown off into the water by small stationary fins or plows which are in such a position that they catch the pieces nearest the edges and force them off as the plate revolves. This arrangement, while allowing a free passage for the carbide, prevents an excess from falling should the machine stop in any position.

When, as is usually the case, the feed mechanism is actuated by the rise or fall of pressure in the generator or of the level of some part of the gas holder, it must be built in such a way that the feeding remains inoperative as long as the filling opening on the carbide holder remains open.

The feed of carbide should always be shut off and controlled so that under no condition can more gas be generated than could be cared for by the relief valve provided. It is necessary also to have the feed mechanism at least ten inches above the surface of the water so that the parts will never become clogged with damp lime dust.

Motor Feed.—The feed mechanism itself is usually operated by power secured from a slowly falling weight which, through a cable, revolves a drum. To this drum is attached suitable gearing for moving the feed parts with sufficient power and in the way desired. This part, called the motor, is controlled by two levers, one releasing a brake and allowing the motor to operate the feed, the other locking the gearing so that no more carbide will be dropped into the water. These levers are moved either by the quantity of gas in the holder or by the pressure of the gas, depending on the type of machine.

With a separate gas holder, such as used with low pressure systems, the levers are operated by the rise and fall of the bell of the holder or gasometer, alternately starting and stopping the motor as the bell falls and rises again. Medium pressure generators are provided with a diaphragm to control the feed motor.

This diaphragm is carried so that the pressure within the generator acts on one side while a spring, whose tension is under the control of the operator, acts on the other side. The diaphragm is connected to the brake and locking device on the motor in such a way that increasing the tension on the spring presses the diaphragm and moves a rod that releases the brake and starts the feed. The gas pressure, increasing with the continuation of carbide feed, acts on the other side and finally overcomes the pressure of the spring tension, moving the control rod the other way and stopping the motor and carbide feed. This spring tension is adjusted and checked with the help of a pressure gauge attached to the generating chamber.

Gravity Feed.—This type of feed differs from the foregoing in that the carbide is simply released and is allowed to fall into the water without being forced to do so. Any form of valve that is sufficiently powerful in action to close with the carbide passing through is used and is operated by the power secured from the rise and fall of the gas holder bell. When this valve is first opened the carbide runs into the water until sufficient pressure and volume of gas is generated to raise the bell. This movement operates the arm attached to the carbide shut off valve and slowly closes it. A fall of the bell occasioned by gas being withdrawn again opens the valve and more gas is generated.

Mechanical Feed.—The previously described methods of feeding carbide to the water have all been automatic in action and do not depend on the operator for their proper action.

Some types of large generating plants have a power-driven feed, the power usually being from some kind of motor other than one operated by a weight, such as a water motor, for instance. This motor is started and stopped by the operator when, in his judgment, more gas is wanted or enough has been generated. This type of machine, often called a "non-automatic generator," is suitable for large installations and is attached to a gas holder of sufficient size to hold a day's supply of acetylene. The generator can then be operated until a quantity of gas has been made that will fill the large holder, or gasometer, and then allowed to remain idle for some time.

Gas Holders. — The commonest type of gas container is that known as a gasometer. This consists of a circular tank partly filled with water, into which is lowered another circular tank, inverted, which is made enough smaller in diameter than the first one so that three-quarters of an inch is left between them. This upper and inverted portion, called the bell, receives the gas from the generator and rises or falls in the bath of water provided in the lower tank as a greater or less amount of gas is contained in it.

These holders are made large enough so that they will provide a means of caring for any after generation and so that they maintain a steady and even flow. The generator, however, must be of a capacity great enough so that the gas holder will not be drawn on for part of the supply with all torches in operation. That is, the holder must not be depended on for a reserve supply.

The bell of the holder is made so that when full of gas its lower edge is still under a depth of at least nine inches of water in the lower tank. Any further rise beyond this point should always release the gas, or at least part of it, to the escape pipe so that the gas will under no circumstances be forced into the room from, between the bell and tank. The bell is guided in its rise and fall by vertical rods so that it will not wedge at any point in its travel.

A condensing chamber to receive the water which condenses from the acetylene gas in the holder is usually placed under this part and is provided with a drain so that this water of condensation may be easily removed.

Filtering. — A small chamber containing some closely packed but porous material such as felt is placed in the pipe leading to the torch lines. As the acetylene gas passes through this filter the particles of lime dust and other impurities are extracted from it so that danger of clogging the torch openings is avoided as much as possible.

The gas is also filtered to a large extent by its passage through the water in the generating chamber, this filtering or "scrubbing" often being facilitated by the form of piping through which the gas must pass from the generating chamber into the holder. If the gas passes out of a number of small openings when going into the holder the small bubbles give a better washing than large ones would.

Piping.—Connections from generators to service pipes should preferably be made with right and left couplings or long thread nipples with lock nuts. If unions are used, they should be of a type that does not require gaskets. The piping should be carried and supported so that any moisture condensing in the lines will drain back toward the generator and where low points occur they should be drained through tees leading into drip cups which are permanently closed with screw caps or plugs. No pet cocks should be used for this purpose.

For the feed pipes to the torch lines the following pipe sizes are recommended.

3/8 inch pipe. 26 feet long. 2 cubic feet per hour.
1/2 inch pipe. 30 feet long. 4 cubic feet per hour.
3/4 inch pipe. 50 feet long. 15 cubic feet per hour.
1 inch pipe. 70 feet long. 27 cubic feet per hour.
1-1/4 inch pipe. 100 feet long. 50 cubic feet per hour.
1-1/2 inch pipe. 150 feet long. 65 cubic feet per hour.
2 inch pipe. 200 feet long. 125 cubic feet per hour.
2-1/2 inch pipe. 300 feet long. 190 cubic feet per hour.
3 inch pipe. 450 feet long. 335 cubic feet per hour.

When drainage is possible into a sewer, the generator should not be connected directly into the sewer but should first discharge into an open receptacle, which may in turn be connected to the sewer.

No valves or pet cocks should open into the generator room or any other room when it would be possible, by opening them for draining purposes, to allow any escape of gas. Any condensation must be removed without the use of valves or other working parts, being drained into closed receptacles. It should be needless to say that all the piping for gas must be perfectly tight at every point in its length.

Safety Devices.—Good generators are built in such a way that the operator must follow the proper order of operation in charging and cleaning as well as in all other necessary care. It has been mentioned that the gas pressure is released or shut off before it is possible to fill the water compartment, and this same idea is carried further in making the generator inoperative and free from gas pressure before opening the residue drain of the carbide filling opening on top of the hopper. Some machines are made so that they automatically cease to generate should there be a sudden and abnormal withdrawal of gas such as would be caused by a bad leak. This method of adding safety by automatic means and interlocking parts may be carried to any extent that seems desirable or necessary to the maker.

All generators should be provided with escape or relief pipes of large size which lead to the open air. These pipes are carried so that condensation will drain back toward the generator and after being led out of the building to a point at least twelve feet above ground, they end in a protecting hood so that no rain or solid matter can find its way into them. Any escape of gas which might ordinarily pass into the generator room is led into these escape pipes, all parts of the system being connected with the pipe so that the gas will find this way out.

Safety blow off valves are provided so that any excess gas which cannot be contained by the gas holder may be allowed to escape without causing an undue rise in pressure. This valve also allows the escape of pressure above that for which the generator was designed. Gas released in this way passes into the escape pipe just described.

Inasmuch as the pressure of the oxygen is much greater than that of the acetylene when used in the torch, it will be seen that anything that caused the torch outlet to become closed would allow the oxygen to force the acetylene back into the generator and the oxygen would follow it, making a very explosive mixture. This return of the gas is prevented by a hydraulic safety valve or back pressure valve, as it is often called.

Mechanical check valves have been found unsuitable for this use and those which employ water as a seal are now required by the insurance rules. The valve itself (Figure 13) consists of a large cylinder containing water to a certain depth, which is indicated on the valve body. Two pipes come into the upper end of this cylinder and lead down into the water, one being longer than the other. The shorter pipe leads to the escape pipe mentioned above, while the longer one comes from the generator. The upper end of the cylinder has an opening to which is attached the pipe leading to the torches.

[Illustration: Figure 13.—Hydraulic Back-Pressure Valve. *A*, Acetylene supply line; *B*, Vent pipe; *C*, Water filling plug; *D*, Acetylene service cock; *E*, Plug to gauge height of water; *F*, Gas openings under water; *G*, Return pipe for sealing water; *H*, Tube to carry gas below water line; *I*, Tube to carry gas to escape pipe; *J*, Gas chamber; *K*, Plug in upper gas chamber; *L*, High water level; *M*, Opening through which water returns; *O*, Bottom clean out casting]

The gas coming from the generator through the longer pipe passes out of the lower end of the pipe which is under water and bubbles up through the water to the space in the top of the cylinder. From there the gas goes to the pipe leading to the torches. The shorter pipe is closed by the depth of water so that the gas does not escape to the relief pipe. As long as the gas flows in the normal direction as described there will be no escape to the air. Should the gas in the torch line return into the hydraulic valve its pressure will lower the level of water in the cylinder by forcing some of the liquid up into the two pipes. As the level of the water lowers, the shorter pipe will be uncovered first, and as this is the pipe leading to the open air the gas will be allowed to escape, while the pipe leading back to the generator is still closed by the water seal. As soon as this reverse flow ceases, the water will again resume its level and the action will continue. Because of the small amount of water blown out of the escape pipe each time the valve is called upon to perform this duty, it is necessary to see that the correct water level is always maintained.

While there are modifications of this construction, the same principle is used in all types. The pressure escape valve is often attached to this hydraulic valve body.

Construction Details.—Flexible tubing (except at torches), swing pipe joints, springs, mechanical check valves, chains, pulleys and lead or fusible piping should never be used on acetylene apparatus except where the failure of those parts will not affect the safety of the machine or permit, either directly or indirectly, the escape of gas into a room. Floats should not be used except where failure will only render the machine inoperative.

It should be said that the National Board of Fire Underwriters have established an inspection service for acetylene generators and any apparatus which bears their label, stating that that particular model and type has been passed, is safe to use. This service is for the best interests of all concerned and looks toward the prevention of accidents. Such inspection is a very important and desirable feature of any outfit and should be insisted upon.

Location of Generators.—Generators should preferably be placed outside of insured buildings and in properly constructed generator houses. The operating mechanism should have ample room to work in and there should be room enough for the attendant to reach the various parts and perform the required duties without hindrance or the need of artificial light. They should also be protected from tampering by unauthorized persons.

Generator houses should not be within five feet of any opening into, nor have any opening toward, any adjacent building, and should be kept under lock and key. The size of the house should be no greater than called for by the requirements mentioned above and it should be well ventilated.

The foundation for the generator itself should be of brick, stone, concrete or iron, if possible. If of wood, they should be extra heavy, located in a dry place and open to circulation of air. A board platform is not satisfactory, but the foundation should be of heavy planking or timber to make a firm base and so that the air can circulate around the wood.

The generator should stand level and no strain should be placed on any of the pipes or connections or any parts of the generator proper.

CHAPTER IV

WELDING INSTRUMENTS

VALVES

Tank Valves.—The acetylene tank valve is of the needle type, fitted with suitable stuffing box nuts and ending in an exposed square shank to which the special wrench may be fitted when the valve is to be opened or closed.

The valve used on Linde oxygen cylinders is also a needle type, but of slightly more complex construction. The body of the valve, which screws into the top of the cylinder, has an opening below through which the gas comes from the cylinder, and another opening on the side through which it issues to the torch line. A needle screws down from above to close this lower opening. The needle which closes the valve is not connected directly to the threaded member, but fits loosely into it. The threaded part is turned by a small hand wheel attached to the upper end. When this hand wheel is turned to the left, or up, as far as it will go, opening the valve, a rubber disc is compressed inside of the valve body and this disc serves to prevent leakage of the gas around the spindle.

The oxygen valve also includes a safety nut having a small hole through it closed by a fusible metal which melts at 250° Fahrenheit. Melting of this plug allows the gas to exert its pressure against a thin copper diaphragm, this diaphragm bursting under the gas pressure and allowing the oxygen to escape into the air.

The hand wheel and upper end of the valve mechanism are protected during shipment by a large steel cap which covers them when screwed on to the end of the cylinder. This cap should always be in place when tanks are received from the makers or returned to them.

[Illustration: Figure 14.—Regulating Valve]

Regulating Valves. — While the pressure in the gas containers may be anything from zero to 1,800 pounds, and will vary as the gas is withdrawn, the pressure of the gas admitted to the torch must be held steady and at a definite point. This is accomplished by various forms of automatic regulating valves, which, while they differ somewhat in details of construction, all operate on the same principle.

The regulator body (Figure 14) carries a union which attaches to the side outlet on the oxygen tank valve. The gas passes through this union, following an opening which leads to a large gauge which registers the pressure on the oxygen remaining in the tank and also to a very small opening in the end of a tube. The gas passes through this opening and into the interior of the regulator body. Inside of the body is a metal or rubber diaphragm placed so that the pressure of the incoming gas causes it to bulge slightly. Attached to the diaphragm is a sleeve or an arm tipped with a small piece of fibre, the fibre being placed so that it is directly opposite the small hole through which the gas entered the diaphragm chamber. The slight movement of the diaphragm draws the fibre tightly over the small opening through which the gas is entering, with the result that further flow is prevented.

Against the opposite side of the diaphragm is the end of a plunger. This plunger is pressed against the diaphragm by a coiled spring. The tension on the coiled spring is controlled by the operator through a threaded spindle ending in a wing or milled nut on the outside of the regulator body. Screwing in on the nut causes the tension on the spring to increase, with a consequent increase of pressure on the side of the diaphragm opposite to that on which the gas acts. Inasmuch as the gas pressure acted to close the small gas opening and the spring pressure acts in the opposite direction from the gas, it will be seen that the spring pressure tends to keep the valve open.

When the nut is turned way out there is of course, no pressure on the spring side of the diaphragm and the first gas coming through automatically closes the opening through which it entered. If now the tension on the spring be slightly increased, the valve will again open and admit gas until the pressure of gas within the regulator is just sufficient to overcome the spring pressure and again close the opening. There will then be a pressure of gas within the regulator that corresponds to the pressure placed on the spring by the operator. An opening leads from the regulator interior to the torch lines so that all gas going to the torches is drawn from the diaphragm chamber.

Any withdrawal of gas will, of course, lower the pressure of that remaining inside the regulator. The spring tension, remaining at the point determined by the operator, will overcome this lessened pressure of the gas, and the valve will again open and admit enough more gas to bring the pressure back to the starting point. This action continues as long as the spring tension remains at this point and as long as any gas is taken from the regulator. Increasing the spring tension will require a greater gas pressure to close the valve and the pressure of that in the regulator will be correspondingly higher.

When the regulator is not being used, the hand nut should be unscrewed until no tension remains on the spring, thus closing the valve. After the oxygen tank valve is open, the regulator hand nut is slowly screwed in until the spring tension is sufficient to give the required pressure in the torch lines. Another gauge is attached to the regulator so that it communicates with the interior of the diaphragm chamber, this gauge showing the gas pressure going to the torch. It is customary to incorporate a safety valve in the regulator which will blow off at a dangerous pressure.

In regulating valves and tank valves, as well as all other parts with which the oxygen comes in contact, it is not permissible to use any form of oil or grease because of danger of ignition and explosion. The mechanism of a regulator is too delicate to be handled in the ordinary shop and should any trouble or leakage develop in this part of the equipment it should be sent to a company familiar with this class of work for the necessary repairs. Gas must never be admitted to a regulator until the hand nut is all the way out, because of danger to the regulator itself and to the operator as well. A regulator can only be properly adjusted when the tank valve and torch valves are fully opened.

[Illustration: Figure 15.—High and Low Pressure Gauges with Regulator]

Acetylene regulators are used in connection with tanks of compressed gas. They are built on exactly the same lines as the oxygen regulating valve and operate in a similar way. One gauge only, the low pressure indicator, is used for acetylene regulators, although both high and low pressure may be used if desired. (See Figure 15.)

TORCHES

Flame is always produced by the combustion of a gas with oxygen and in no other way. When we burn oil or candles or anything else, the material of the fuel is first turned to a gas by the heat and is then burned by combining with the oxygen of the air. If more than a normal supply of air is forced into the flame, a greater heat and more active burning follows. If the amount of air, and consequently oxygen, is reduced, the flame becomes smaller and weaker and the combustion is less rapid. A flame may be easily extinguished by shutting off all of its air supply.

The oxygen of the combustion only forms one-fifth of the total volume of air; therefore, if we were to supply pure oxygen in place of air, and in equal volume, the action would be several times as intense. If the oxygen is mixed with the fuel gas in the proportion that burns to the very best advantage, the flame is still further strengthened and still more heat is developed because of the perfect combustion. The greater the amount of fuel gas that can be burned in a certain space and within a certain time, the more heat will be developed from that fuel.

The great amount of heat contained in acetylene gas, greater than that found in any other gaseous fuel, is used by leading this gas to the oxy-acetylene torch and there combining it with just the right amount of oxygen to make a flame of the greatest power and heat than can possibly be produced by any form of combustion of fuels of this kind. The heat developed by the flame is about 6300° Fahrenheit and easily melts all the metals, as well as other solids.

Other gases have been and are now being used in the torch. None of them, however, produce the heat that acetylene does, and therefore the oxy-acetylene process has proved the most useful of all. Hydrogen was used for many years before acetylene was introduced in this field. The oxy-hydrogen flame develops a heat far below that of oxy-acetylene, namely 4500° Fahrenheit. Coal gas, benzine gas, blaugas and others have also been used in successful applications, but for the present we will deal exclusively with the acetylene fuel.

It was only with great difficulty that the obstacles in the way of successfully using acetylene were overcome by the development of practicable controlling devices and torches, as well as generators. At present the oxy-acetylene process is the most universally adaptable, and probably finds the most widely extended field of usefulness of any welding process.

The theoretical proportion of the gases for perfect combustion is two and one-half volumes of oxygen to one of acetylene. In practice this proportion is one and one-eighth or one and one-quarter volumes of oxygen to one volume of acetylene, so that the cost is considerably reduced below what it would be if the theoretical quantity were really necessary, as oxygen costs much more than acetylene in all cases.

While the heat is so intense as to fuse anything brought into the path of the flame, it is localized in the small "welding cone" at the torch tip so that the torch is not at all difficult to handle without special protection except for the eyes, as already noted. The art of successful welding may be acquired by any operator of average intelligence within a reasonable time and with some practice. One trouble met with in the adoption of this process has been that the operation looks so simple and so easy of performance that unskilled and unprepared persons have been tempted to try welding, with results that often caused condemnation of the process, when the real fault lay entirely with the operator.

The form of torch usually employed is from twelve to twenty-four inches long and is composed of a handle at one end with tubes leading from this handle to the "welding head" or torch proper. At or near one end of the handle are adjustable cocks or valves for allowing the gases to flow into the torch or to prevent them from doing so. These cocks are often used for regulating the pressure and amount of gas flowing to the welding head, but are not always constructed for this purpose and should not be so used when it is possible to secure pressure adjustment at the regulators (Figure 16).

Figure 16 shows three different sizes of torches. The number 5 torch is designed especially for jewelers' work and thin sheet steel welding. It is eleven inches in length and weighs nineteen ounces. The tips for the number 10 torch are interchangeable with the number 5. The number 10 torch is adapted for general use on light and medium heavy work. It has six tips and its length is sixteen inches, with a weight of twenty-three ounces.

The number 15 torch is designed for heavy work, being twenty-five inches in length, permitting the operator to stand away from the heat of the metal being worked. These heavy tips are in two parts, the oxygen check being renewable.

[Illustration: Figure 16. — Three Sizes of Torches, with Tips]

Figures 17 and 18 show two sizes of another welding torch. Still another type is shown in Figure 19 with four interchangeable tips, the function of each being as follows:

No. 1. For heavy castings.
No. 2. Light castings and heavy sheet metal.
No. 3. Light sheet metal.
No. 4. Very light sheet metal and wire.

[Illustration: Figure 17. — Cox Welding Torch (No. 1)]

[Illustration: Figure 18. — Cox Welding Torch (No. 2)]

[Illustration: Figure 19. — Monarch Welding Torch]

At the side of the shut off cock away from the torch handle the gas tubes end in standard forms of hose nozzles, to which the rubber hose from the gas supply tanks or generators can be attached. The tubes from the handle to the head may be entirely separate from each other, or one may be contained within the other. As a general rule the upper one of two separate tubes carries the oxygen, while this gas is carried in the inside tube when they are concentric with each other.

In the welding head is the mixing chamber designed to produce an intimate mixture of the two gases before they issue from the nozzle to the flame. The nozzle, or welding tip, of a suitable size are design for the work to be handled and the pressure of gases being used, is attached to the welding head and consists essentially of the passage at the outer end of which the flame appears.

The torch body and tubes are usually made of brass, although copper is sometimes used. The joint must be very strong, and are usually threaded and soldered with silver solder. The nozzle proper is made from copper, because it withstands the heat of the flame better than other less suitable metals. The torch must be built in such a way that it is not at all liable to come apart under the influence of high temperatures.

All torches are constructed in such a way that it is impossible for the gases to mix by any possible chance before they reach the head, and the amount of gas contained in the head and tip after being mixed is made as small as possible. In order to prevent the return of the flame through the acetylene tube under the influence of the high pressure oxygen some form of back flash preventer is usually incorporated in the torch at or near the point at which the acetylene enters. This preventer takes the form of some porous and heat absorbing material, such as aluminum shavings, contained in a small cavity through which the gas passes on its way to the head.

High Pressure Torches.—Torches are divided into the same classes as are the generators; that is, high pressure, medium pressure and low pressure. As mentioned before, the medium pressure is usually called the high pressure, because there are very few true high pressure systems in use, and comparatively speaking the medium pressure type is one of high pressure.

[Illustration: Figure 20.—High Pressure Torch Head]

With a true high pressure torch (Figure 20) the gases are used at very nearly equal heads so that the mixing before ignition is a simple matter. This type admits the oxygen at the inner end of a straight passage leading to the tip of the nozzle. The acetylene comes into this same passage from openings at one side and near the inner end. The difference in direction of the two gases as they enter the passage assists in making a homogeneous mixture. The construction of this nozzle is perfectly simple and is easily understood. The true high pressure torch nozzle is only suited for use with compressed and dissolved acetylene, no other gas being at a sufficient pressure to make the action necessary in mixing the gases.

Medium Pressure Torches.—The medium pressure (usually called high pressure) torch (Figure 21) uses acetylene from a medium pressure generator or from tanks of compressed gas, but will not take the acetylene from low pressure generators.

[Illustration: Figure 21.—Medium Pressure Torch Head]

The construction of the mixing chamber and nozzle is very similar to that of the high pressure torch, the gases entering in the same way and from the same positions of openings. The pressure of the acetylene is but little lower than that of the oxygen, and the two gases, meeting at right angles, form a very intimate mixture at this point of juncture. The mixture in its proportions of gases depends entirely on the sizes of the oxygen and acetylene openings into the mixing chamber and on the pressures at which the gases are admitted. There is a very slight injector action as the fast moving stream of oxygen tends to draw the acetylene from the side openings into the chamber, but the operation of the torch does not depend on this action to any extent.

Low Pressure Torches.—The low pressure torch (Figure 22) will use gas from low pressure generators from medium pressure machines or from tanks in which it has been compressed and dissolved. This type depends for a perfect mixture of gas upon the principle of the injector just as it is applied in steam boiler practice.

[Illustration: Figure 22.—Low Pressure Torch with Separate Injector Nozzle]

The oxygen enters the head at considerable pressure and passes through its tube to a small jet within the head. The opening of this jet is directly opposite the end of the opening through the nozzle which forms the mixing chamber and the path of the gases to the flame. A small distance remains between the opening from which the oxygen issues and the inner opening into the mixing passage. The stream of oxygen rushes across this space and enters the mixing chamber, being driven by its own pressure.

The acetylene enters the head in an annular space surrounding the oxygen tube. The space between oxygen jet and mixing chamber opening is at one end of this acetylene space and the stream of oxygen seizes the acetylene and under the injector action draws it into the mixing chamber, it being necessary only to have a sufficient supply of acetylene flowing into the head to allow the oxygen to draw the required proportion for a proper mixture.

The volume of gas drawn into the mixing chamber depends on the size of the injector openings and the pressure of the oxygen. In practice the oxygen pressure is not altered to produce different sized flames, but a new nozzle is substituted which is designed to give the required flame. Each nozzle carries its own injector, so that the design is always suited to the conditions. While torches are made having the injector as a permanent part of the torch body, the replaceable nozzle is more commonly used because it makes the one torch suitable for a large range of work and a large number of different sized flames. With the replaceable head a definite pressure of oxygen is required for the size being used, this pressure being the one for which the injector and corresponding mixing chamber were designed in producing the correct mixture.

Adjustable Injectors.-Another form of low pressure torch operates on the injector principle, but the injector itself is a permanent part of the torch, the nozzle only being changed for different sizes of work and flame. The injector is placed in or near the handle and its opening is the largest required by any work that can be handled by this particular torch. The opening through the tip of the injector through which the oxygen issues on its way to the mixing chamber may be wholly or partly closed by a needle valve which may be screwed into the opening or withdrawn from it, according to the operator's judgment. The needle valve ends in a milled nut outside the torch handle, this being the adjustment provided for the different nozzles.

Torch Construction.—A well designed torch is so designed that the weight distribution is best for holding it in the proper position for welding. When a torch is grasped by its handle with the gas hose attached, it should balance so that it does not feel appreciably heavier on one end than on the other.

The head and nozzle may be placed so that the flame issues in a line at right angles with the torch body, or they may be attached at an angle convenient for the work to be done. The head set at an angle of from 120 to 170 degrees with the body is usually preferred for general work in welding, while the cutting torch usually has its head at right angles to the body.

Removable nozzles have various size openings through them and the different sizes are designated by numbers from 1 up. The same number does not always indicate the same size opening in torches of different makes, nor does it indicate a nozzle of the same capacity.

The design of the nozzle, the mixing chamber, the injector, when one is used, and the size of the gas openings must be such that all these things are suited to each other if a proper mixture of gas is to be secured. Parts that are not made to work together are unsafe if used because of the danger of a flash back of the flame into the mixing chamber and gas tubes. It is well known that flame travels through any inflammable gas at a certain definite rate of speed, depending on the degree of inflammability of the gas. The easier and quicker the gas burns, the faster will the flame travel through it.

If the gas in the nozzle and mixing chamber stood still, the flame would immediately travel back into these parts and produce an explosion of more or less violence. The speed with which the gases issue from the nozzle prevent this from happening because the flame travels back through the gas at the same speed at which the gas issues from the torch tip. Should the velocity of the gas be greater than the speed of flame propagation through it, it will be impossible to keep the flame at the tip, the tendency being for a space of unburned gas to appear between tip and flame. On the other hand, should the speed of the flame exceed the velocity with which the gas comes from the torch there will result a flash back and explosion.

Care of Torches.—An oxy-acetylene torch is a very delicate and sensitive device, much more so that appears on the surface. It must be given equally as good care and attention as any other high-priced piece of machinery if it is to be maintained in good condition for use.

It requires cleaning of the nozzles at regular intervals if used regularly. This cleaning is accomplished with a piece of copper or brass wire run through the opening, and never with any metal such as steel or iron that is harder than the nozzle itself, because of the danger of changing the size of the openings. The torch head and nozzle can often be cleaned by allowing the oxygen to blow through at high pressure without the use of any tools.

In using a torch a deposit of carbon will gradually form inside of the head, and this deposit will be more rapid if the operator lights the stream of acetylene before turning any oxygen into the torch. This deposit may be removed by running kerosene through the nozzle while it is removed from the torch, setting fire to the kerosene and allowing oxygen to flow through while the oil is burning.

Should a torch become clogged in the head or tubes, it may usually be cleaned by removing the oxygen hose from the handle end, closing the acetylene cock on the torch, placing the end of the oxygen hose over the opening in the nozzle and turning on the oxygen under pressure to blow the obstruction back through the passage that it has entered. By opening the acetylene cock and closing the oxygen cock at the handle, the acetylene passages may then be cleaned in the same way. Under no conditions should a torch be taken apart any more than to remove the changeable nozzle, except in the hands of those experienced in this work.

Nozzle Sizes.—The size of opening through the nozzle is determined according to the thickness and kind of metal being handled. The following sizes are recommended for steel:

Davis-Bourn

onville. Oxweld Low Thickness of Metal

(Medium Pressure.) Pressure 1/32 Tip No

Head No.	1	2	3
1	1/16	5/64	3/32
2			3
3			4
			3

/8
4
5
3/16
5
6
1/4
6
7
5/16
7
3/8
8
8

1/2

9

10

5/8

10

12

3/4

11

15

Very

heavy

$$\frac{1}{2}$$

$$\frac{1}{5}$$

Cutting Torches.—Steel may be cut with a jet of oxygen at a rate of speed greater than in any other practicable way under usual conditions. The action consists of burning away a thin section of the metal by allowing a stream of oxygen to flow onto it while the gas is at high pressure and the metal at a white heat.

[Illustration: Figure 23.—Cutting Torch]

The cutting torch (Figure 23) has the same characteristics as the welding torch, but has an additional nozzle or means for temporarily using the welding opening for the high pressure oxygen. The oxygen issues from the opening while cutting at a pressure of from ten to 100 pounds to the square inch.

The work is first heated to a white heat by adjusting the torch for a welding flame. As soon as the metal reaches this temperature, the high pressure oxygen is turned on to the white-hot portion of the steel. When the jet of gas strikes the metal it cuts straight through, leaving a very narrow slot and removing but little metal. Thicknesses of steel up to ten inches can be economically handled in this way.

The oxygen nozzle is usually arranged so that it is surrounded by a number of small jets for the heating flame. It will be seen that this arrangement makes the heating flame always precede the oxygen jet, no matter in which direction the torch is moved.

The torch is held firmly, either by hand or with the help of special mechanism for guiding it in the desired path, and is steadily advanced in the direction it is desired to extend the cut, the rate of advance being from three inches to two feet per minute through metal from nine inches down to one-quarter of an inch in thickness.

The following data on cutting is given by the Davis-Bournonville Company:

Cubic Feet Cost of Thickness of Gas Inches Gases of Cutting Heating per Foot Oxygen Cut per per Foot Steel Oxygen Oxygen of Cut Acetylene Min. of Cut 1/4 10 lbs. 4 lbs. .40 .086 24 $ .013 1/2 20 4 .91 .150 15 .029 3/4 30 4 1.16 .150 15 .036 1 30 4 1.45 .172 12 .045 1 1/2 30 5 2.40 .380 12 .076 2 40 5 2.96 .380 12 .093 4 50 5 9.70 .800 7 .299 6 70 6 21.09 1.50 4 .648 9 100 6 43.20 2.00 3 1.311

Acetylene-Air Torch.—A form of torch which burns the acetylene after mixing it with atmospheric air at normal pressure rather than with the oxygen under higher pressures has been found useful in certain pre-heating, brazing and similar operations. This torch (Figure 24) is attached by a rubber gas hose to any compressed acetylene tank and is regulated as to flame size and temperature by opening or closing the tank valve more or less.

After attaching the torch to the tank, the gas is turned on very slowly and is lighted at the torch tip. The adjustment should cause the presence of a greenish-white cone of flame surrounded by a larger body of burning gas, the cone starting at the mouth of the torch.

[Illustration: Figure 24.—Acetylene-Air Torch]

By opening the tank valve more, a longer and hotter flame is produced, the length being regulated by the tank valve also. This torch will give sufficient heat to melt steel, although not under conditions suited to welding. Because of the excess of acetylene always present there is no danger of oxidizing the metal being heated.

The only care required by this torch is to keep the small air passages at the nozzle clean and free from carbon deposits. The flame should be extinguished when not in use rather than turned low, because this low flame rapidly deposits large quantities of soot in the burner.

CHAPTER V

OXY-ACETYLENE WELDING PRACTICE

PREPARATION OF WORK

Preheating.—The practice of heating the metal around the weld before applying the torch flame is a desirable one for two reasons. First, it makes the whole process more economical; second, it avoids the danger of breakage through expansion and contraction of the work as it is heated and as it cools.

When it is desired to join two surfaces by welding them, it is, of course, necessary to raise the metal from the temperature of the surrounding air to its melting point, involving an increase in temperature of from one thousand to nearly three thousand degrees. To obtain this entire increase of temperature with the torch flame is very wasteful of fuel and of the operator's time. The total amount of heat necessary to put into metal is increased by the conductivity of that metal because the heat applied at the weld is carried to other parts of the piece being handled until the whole mass is considerably raised in temperature. To secure this widely distributed increase the various methods of preheating are adopted.

As to the second reason for preliminary heating. It is understood that the metal added to the joint is molten at the time it flows into place. All the metals used in welding contract as they cool and occupy a much smaller space than when molten. If additional metal is run between two adjoining surfaces which are parts of a surrounding body of cool metal, this added metal will cool while the surfaces themselves are held stationary in the position they originally occupied. The inevitable result is that the metal added will crack under the strain, or, if the weld is exceptionally strong, the main body of the work will he broken by the force of contraction. To overcome these difficulties is the second and most important reason for preheating and also for slow cooling following the completion of the weld.

There are many ways of securing this preheating. The work may be brought to a red heat in the forge if it is cast iron or steel; it may he heated in special ovens built for the purpose; it may be placed in a bed of charcoal while suitably supported; it may be heated by gas or gasoline preheating torches, and with very small work the outer flame of the welding torch automatically provides means to this end.

The temperature of the parts heated should be gradually raised in all cases, giving the entire mass of metal a chance to expand equally and to adjust itself to the strains imposed by the preheating. After the region around the weld has been brought to a proper temperature the opening to be filled is exposed so that the torch flame can reach it, while the remaining surfaces are still protected from cold air currents and from cooling through natural radiation.

One of the commonest methods and one of the best for handling work of rather large size is to place the piece to be welded on a bed of fire brick and build a loose wall around it with other fire brick placed in rows, one on top of the other, with air spaces left between adjacent bricks in each row. The space between the brick retaining wall and the work is filled with charcoal, which is lighted from below. The top opening of the temporary oven is then covered with asbestos and the fire kept up until the work has been uniformly raised in temperature to the desired point.

When much work of the same general character and size is to be handled, a permanent oven may be constructed of fire brick, leaving a large opening through the top and also through one side. Charcoal may be used in this form of oven as with the temporary arrangement, or the heat may be secured from any form of burner or torch giving a large volume of flame. In any method employing flame to do the heating, the work itself must be protected from the direct blast of the fire. Baffles of brick or metal should be placed between the mouth of the torch and the nearest surface of the work so that the flame will be deflected to either side and around the piece being heated.

The heat should be applied to bring the point of welding to the highest temperature desired and, except in the smallest work, the heat should gradually shade off from this point to the other parts of the piece. In the case of cast iron and steel the temperature at the point to be welded should be great enough to produce a dull red heat. This will make the whole operation much easier, because there will be no surrounding cool metal to reduce the temperature of the molten material from the welding rod below the point at which it will join the work. From this red heat the mass of metal should grow cooler as the distance from the weld becomes greater, so that no great strain is placed upon any one part. With work of a very irregular shape it is always best to heat the entire piece so that the strains will be so evenly distributed that they can cause no distortion or breakage under any conditions.

The melting point of the work which is being preheated should be kept in mind and care exercised not to approach it too closely. Special care is necessary with aluminum in this respect, because of its low melting temperature and the sudden weakening and flowing without warning. Workmen have carelessly overheated aluminum castings and, upon uncovering the piece to make the weld, have been astonished to find that it had disappeared. Six hundred degrees is about the safe limit for this metal. It is possible to gauge the exact temperature of the work with a pyrometer, but when this instrument cannot be procured, it might be well to secure a number of "temperature cones" from a chemical or laboratory supply house. These cones are made from material that will soften at a certain heat and in form they are long and pointed. Placed in position on the part being heated, the point may be watched, and when it bends over it is sure that the metal itself has reached a temperature considerably in excess of the temperature at which that particular cone was designed to soften.

The object in preheating the metal around the weld is to cause it to expand sufficiently to open the crack a distance equal to the contraction when cooling from the melting point. In the case of a crack running from the edge of a piece into the body or of a crack wholly within the body, it is usually satisfactory to heat the metal at each end of the opening. This will cause the whole length of the crack to open sufficiently to receive the molten material from the rod.

The judgment of the operator will be called upon to decide just where a piece of metal should be heated to open the weld properly. It is often possible to apply the preheating flame to a point some distance from the point of work if the parts are so connected that the expansion of the heated part will serve to draw the edges of the weld apart. Whatever part of the work is heated to cause expansion and separation, this part must remain hot during the entire time of welding and must then cool slowly at the same time as the metal in the weld cools.

[Illustration: Figure 25.—Preheating at A While Welding at B. C also May Be Heated.]

An example of heating points away from the crack might be found in welding a lattice work with one of the bars cracked through (Figure 25). If the strips parallel and near to the broken bar are heated gradually, the work will be so expanded that the edges of the break are drawn apart and the weld can be successfully made. In this case, the parallel bars next to the broken one would be heated highest, the next row not quite so hot and so on for some distance away. If only the one row were heated, the strains set up in the next ones would be sufficient to cause a new break to appear.

[Illustration: Figure 26.—Cutting Through the Rim of a Wheel (Cut Shown at A)]

If welding is to be done near the central portion of a large piece, the strains will be brought to bear on the parts farthest away from the center. Should a fly wheel spoke be broken and made ready to weld, the greatest strain will come on the rim of the wheel. In cases like this it is often desirable to cut through at the point of greatest strain with a saw or cutting torch, allowing free movement while the weld is made at the original break (Figure 26). After the inside weld is completed, the cut may be welded without danger, for the reason that it will always be at some point at which severe strains cannot be set up by the contraction of the cooling metal.

[Illustration: Figure 27.—Using a Wedge While Welding]

In materials that will spring to some extent without breakage, that is, in parts that are not brittle, it may be possible to force the work out of shape with jacks or wedges (Figure 27) in the same way that it would be distorted by heating and expanding some portion of it as described. A careful examination will show whether this method can be followed in such a way as to force the edges of the break to separate. If the plan seems feasible, the wedges may be put in place and allowed to remain while the weld is completed. As soon as the work is finished the wedges should be removed so that the natural contraction can take place without damage.

It should always be remembered that it is not so much the expansion of the work when heated as it is the contraction caused by cooling that will do the damage. A weld may be made that, to all appearances, is perfect and it may be perfect when completed; but if provision has not been made to allow for the contraction that is certain to follow, there will be a breakage at some point. It is not possible to weld the simplest shapes, other than straight bars, without considering this difficulty and making provision to take care of it.

The exact method to employ in preheating will always call for good judgment on the part of the workman, and he should remember that the success or failure of his work will depend fully as much on proper preparation as on correct handling of the weld itself. It should be remembered that the outer flame of the oxy-acetylene torch may be depended on for a certain amount of preheating, as this flame gives a very large volume of heat, but a heat that is not so intense nor so localized as the welding flame itself. The heat of this part of the flame should be fully utilized during the operation of melting the metal and it should be so directed, when possible, that it will bring the parts next to be joined to as high a temperature as possible.

When the work has been brought to the desired temperature, all parts except the break and the surface immediately surrounding it on both sides should be covered with heavy sheet asbestos. This protecting cover should remain in place throughout the operation and should only be moved a distance sufficient to allow the torch flame to travel in the path of the weld. The use of asbestos in this way serves a twofold purpose. It retains the heat in the work and prevents the breakage that would follow if a draught of air were to strike the heated metal, and it also prevents such a radiation of heat through the surrounding air as would make it almost impossible for the operator to perform his work, especially in the case of large and heavy castings when the amount of heat utilized is large.

Cleaning and Champfering. — A perfect weld can never be made unless the surfaces to be joined have been properly prepared to receive the new metal.

All spoiled, burned, corroded and rough particles must positively be removed with chisel and hammer and with a free application of emery cloth and wire brush. The metal exposed to the welding flame should be perfectly clean and bright all over, or else the additional material will not unite, but will only stick at best.

[Illustration: Figure 28. — Tapering the Opening Formed by a Break]

Following the cleaning it is always necessary to bevel, or champfer, the edges except in the thinnest sheet metal. To make a weld that will hold, the metal must be made into one piece, without holes or unfilled portions at any point, and must be solid from inside to outside. This can only be accomplished by starting the addition of metal at one point and gradually building it up until the outside, or top, is reached. With comparatively thin plates the molten metal may be started from the side farthest from the operator and brought through, but with thicker sections the addition is started in the middle and brought flush with one side and then with the other.

It will readily be seen that the molten material cannot be depended upon to flow between the tightly closed surfaces of a crack in a way that can be at all sure to make a true weld. It will be necessary for the operator to reach to the farthest side with the flame and welding rod, and to start the new surfaces there. To allow this, the edges that are to be joined are beveled from one side to the other (Figure 28), so that when placed together in approximately the position they are to occupy they will leave a grooved channel between them with its sides at an angle with each other sufficient in size to allow access to every point of each surface.

[Illustration: Figure 29. — Beveling for Thin Work]

[Illustration: Figure 30. — Beveling for Thick Work]

With work less than one-fourth inch thick, this angle should be forty-five degrees on each piece (Figure 29), so that when they are placed together the extreme edges will meet at the bottom of a groove whose sides are square, or at right angles, to each other. This beveling should be done so that only a thin edge is left where the two parts come together, just enough points in contact to make the alignment easy to hold. With work of a thickness greater than a quarter of an inch, the angle of bevel on each piece may be sixty degrees (Figure 30), so that when placed together the angle included between the sloping sides will also be sixty degrees. If the plate is less than one-eighth of an inch thick the beveling is not necessary, as the edges may be melted all the way through without danger of leaving blowholes at any point.

[Illustration: Figure 31. — Beveling Both Sides of a Thick Piece]

[Illustration: Figure 32. — Beveling the End of a Pipe]

This beveling may be done in any convenient way. A chisel is usually most satisfactory and also quickest. Small sections may be handled by filing, while metal that is too hard to cut in either of these ways may be shaped on the emery wheel. It is not necessary that the edges be perfectly finished and absolutely smooth, but they should be of regular outline and should always taper off to a thin edge so that when the flame is first applied it can be seen issuing from the far side of the crack. If the work is quite thick and is of a shape that will allow it to be turned over, the bevel may be brought from both sides (Figure 31), so that there will be two grooves, one on each surface of the work. After completing the weld on one side, the piece is reversed and finished on the other side. Figure 32 shows the proper beveling for welding pipe. Figure 33 shows how sheet metal may be flanged for welding.

Welding should not be attempted with the edges separated in place of beveled, because it will be found impossible to build up a solid web of new metal from one side clear through to the other by this method. The flame cannot reach the surfaces to make them molten while receiving new material from the rod, and if the flame does not reach them it will only serve to cause a few drops of the metal to join and will surely cause a weak and defective weld.

[Illustration: Figure 33. — Flanging Sheet Metal for Welding]

Supporting Work. — During the operation of welding it is necessary that the work be well supported in the position it should occupy. This may be done with fire brick placed under the pieces in the correct position, or, better still, with some form of clamp. The edges of the crack should touch each other at the point where welding is to start and from there should gradually separate at the rate of about one-fourth inch to the foot. This is done so that the cooling of the molten metal as it is added will draw the edges together by its contraction.

Care must be used to see that the work is supported so that it will maintain the same relative position between the parts as must be present when the work is finished. In this connection it must be remembered that the expansion of the metal when heated may be great enough to cause serious distortion and to provide against this is one of the difficulties to be overcome.

Perfect alignment should be secured between the separate parts that are to be joined and the two edges must be held up so that they will be in the same plane while welding is carried out. If, by any chance, one drops below the other while molten metal is being added, the whole job may have to be undone and done over again. One precaution that is necessary is that of making sure that the clamping or supporting does not in itself pull the work out of shape while melted.

TORCH PRACTICE

[Illustration: Figure 34. — Rotary Movement of Torch in Welding]

The weld is made by bringing the tip of the welding flame to the edges of the metals to be joined. The torch should be held in the right hand and moved slowly along the crack with a rotating motion, traveling in small circles (Figure 34), so that the Welding flame touches first on one side of the crack and then on the other. On large work the motion may be simply back and forth across the crack, advancing regularly as the metal unites. It is usually best to weld toward the operator rather than from him, although this rule is governed by circumstances. The head of the torch should be inclined at an angle of about 60 degrees to the surface of the work.

The torch handle should extend in the same line with the break (Figure 35) and not across it, except when welding very light plates.

[Illustration: Figure 35. — Torch Held in Line with the Break]

If the metal is 1/16 inch or less in thickness it is only necessary to circle along the crack, the metal itself furnishing enough material to complete the weld without additions. Heat both sides evenly until they flow together.

Material thicker than the above requires the addition of more metal of the same or different kind from the welding rod, this rod being held by the left hand. The proper size rod for cast iron is one having a diameter equal to the thickness of metal being welded up to a one-half inch rod, which is the largest used. For steel the rod should be one-half the thickness of the metal being joined up to one-fourth inch rod. As a general rule, better results will be obtained by the use of smaller rods, the very small sizes being twisted together to furnish enough material while retaining the free melting qualities.

[Illustration: Figure 36. — The Welding Rod Should Be Held in the Molten
Metal]

The tip of the rod must at all times be held in contact with the pieces being welded and the flame must be so directed that the two sides of the crack and the end of the rod are melted at the same time (Figure 36). Before anything is added from the rod, the sides of the crack are melted down sufficiently to fill the bottom of the groove and join the two sides. Afterward, as metal comes from the rod in filling the crack, the flame is circled along the joint being made, the rod always following the flame.

[Illustration: Figure 37. — Welding Pieces of Unequal Thickness]

Figure 37 illustrates the welding of pieces of unequal thickness.

Figure 38 illustrates welding at an angle.

The molten metal may be directed as to where it should go by the tip of the welding flame, which has considerable force, but care must be taken not to blow melted metal on to cooler surfaces which it cannot join. If, while welding, a spot appears which does not unite with the weld, it may be handled by heating all around it to a white heat and then immediately welding the bad place.

[Illustration: Figure 38.—Welding at an Angle]

Never stop in the middle of a weld, as it is extremely difficult to continue smoothly when resuming work.

The Flame.—The welding flame must have exactly the right proportions of each gas. If there is too much oxygen, the metal will be burned or oxidized; the presence of too much acetylene carbonizes the metal; that is to say, it adds carbon and makes the work harder. Just the right mixture will neither burn nor carbonize and is said to be a "neutral" flame. The neutral flame, if of the correct size for the work, reduces the metal to a melted condition, not too fluid, and for a width about the same as the thickness of the metal being welded.

When ready to light the torch, after attaching the right tip or head as directed in accordance with the thickness of metal to be handled, it will be necessary to regulate the pressure of gases to secure the neutral flame.

The oxygen will have a pressure of from 2 to 20 pounds, according to the nozzle used. The acetylene will have much less. Even with the compressed gas, the pressure should never exceed 10 pounds for the largest work, and it will usually be from 4 to 6. In low pressure systems, the acetylene will be received at generator pressure. It should first be seen that the hand-screws on the regulators are turned way out so that the springs are free from any tension. It will do no harm if these screws are turned back until they come out of the threads. This must be done with both oxygen and acetylene regulators.

Next, open the valve from the generator, or on the acetylene tank, and carefully note whether there is any odor of escaping gas. Any leakage of this gas must be stopped before going on with the work.

The hand wheel controlling the oxygen cylinder valve should now be turned very slowly to the left as far as it will go, which opens the valve, and it should be borne in mind the pressure that is being released. Turn in the hand screw on the oxygen regulator until the small pressure gauge shows a reading according to the requirements of the nozzle being used. This oxygen regulator adjustment should be made with the cock on the torch open, and after the regulator is thus adjusted the torch cock may be closed.

Open the acetylene cock on the torch and screw in on the acetylene regulator hand-screw until gas commences to come through the torch. Light this flow of acetylene and adjust the regulator screw to the pressure desired, or, if there is no gauge, so that there is a good full flame. With the pressure of acetylene controlled by the type of generator it will only be necessary to open the torch cock.

With the acetylene burning, slowly open the oxygen cock on the torch and allow this gas to join the flame. The flame will turn intensely bright and then blue white. There will be an outer flame from four to eight inches long and from one to three inches thick. Inside of this flame will be two more rather distinctly defined flames. The inner one at the torch tip is very small, and the intermediate one is long and pointed. The oxygen should be turned on until the two inner flames unite into one blue-white cone from one-fourth to one-half inch long and one-eighth to one-fourth inch in diameter. If this single, clearly defined cone does not appear when the oxygen torch cock has been fully opened, turn off some of the acetylene until it does appear.

If too much oxygen is added to the flame, there will still be the central blue-white cone, but it will be smaller and more or less ragged around the edges (Figure 39). When there is just enough oxygen to make the single cone, and when, by turning on more acetylene or by turning off oxygen, two cones are caused to appear, the flame is neutral (Figure 40), and the small blue-white cone is called the welding flame.

[Illustration: Figure 39.—Oxidizing Flame—Too Much Oxygen]

[Illustration: Figure 40.—Neutral Flame]

[Illustration: Figure 41.—Reducing Flame—Showing an Excess of Acetylene]

While welding, test the correctness of the flame adjustment occasionally by turning on more acetylene or by turning off some oxygen until two flames or cones appear. Then regulate as before to secure the single distinct cone. Too much oxygen is not usually so harmful as too much acetylene, except with aluminum. (See Figure 41.) An excessive amount of sparks coming from the weld denotes that there is too much oxygen in the flame. Should the opening in the tip become partly clogged, it will be difficult to secure a neutral flame and the tip should be cleaned with a brass or copper wire — never with iron or steel tools or wire of any kind. While the torch is doing its work, the tip may become excessively hot due to the heat radiated from the molten metal. The tip may be cooled by turning off the acetylene and dipping in water with a slight flow of oxygen through the nozzle to prevent water finding its way into the mixing chamber.

The regulators for cutting are similar to those for welding, except that higher pressures may be handled, and they are fitted with gauges reading up to 200 or 250 pounds pressure.

In welding metals which conduct the heat very rapidly it is necessary to use a much larger nozzle and flame than for metals which have not this property. This peculiarity is found to the greatest extent in copper, aluminum and brass.

Should a hole be blown through the work, it may be closed by withdrawing the flame for a few seconds and then commencing to build additional metal around the edges, working all the way around and finally closing the small opening left at the center with a drop or two from the welding rod.

WELDING VARIOUS METALS

Because of the varying melting points, rates of expansion and contraction, and other peculiarities of different metals, it is necessary to give detailed consideration to the most important ones.

Characteristics of Metals. — The welder should thoroughly understand the peculiarities of the various metals with which he has to deal. The metals and their alloys are described under this heading in the first chapter of this book and a tabulated list of the most important points relating to each metal will be found at the end of the present chapter. All this information should be noted by the operator of a welding installation before commencing actual work.

Because of the nature of welding, the melting point of a metal is of great importance. A metal melting at a low temperature should have more careful treatment to avoid undesired flow than one which melts at a temperature which is relatively high. When two dissimilar metals are to be joined, the one which melts at the higher temperature must be acted upon by the flame first and when it is in a molten condition the heat contained in it will in many cases be sufficient to cause fusion of the lower melting metal and allow them to unite without playing the flame on the lower metal to any great extent.

The heat conductivity bears a very important relation to welding, inasmuch as a metal with a high rate of conductance requires more protection from cooling air currents and heat radiation than one not having this quality to such a marked extent. A metal which conducts heat rapidly will require a larger volume of flame, a larger nozzle, than otherwise, this being necessary to supply the additional heat taken away from the welding point by this conductance.

The relative rates of expansion of the various metals under heat should be understood in order that parts made from such material may have proper preparation to compensate for this expansion and contraction. Parts made from metals having widely varying rates of expansion must have special treatment to allow for this quality, otherwise breakage is sure to occur.

Cast Iron.—All spoiled metal should he cut away and if the work is more than one-eighth inch in thickness the sides of the crack should be beveled to a 45 degree angle, leaving a number of points touching at the bottom of the bevel so that the work may be joined in its original relation.

The entire piece should be preheated in a bricked-up oven or with charcoal placed on the forge, when size does not warrant building a temporary oven. The entire piece should be slowly heated and the portion immediately surrounding the weld should be brought to a dull red. Care should be used that the heat does not warp the metal through application to one part more than the others. After welding, the work should be slowly cooled by covering with ashes, slaked lime, asbestos fibre or some other non-conductor of heat. These precautions are absolutely essential in the case of cast iron.

A neutral flame, from a nozzle proportioned to the thickness of the work, should be held with the point of the blue-white cone about one-eighth inch from the surface of the iron.

A cast iron rod of correct diameter, usually made with an excess of silicon, is used by keeping its end in contact with the molten metal and flowing it into the puddle formed at the point of fusion. Metal should be added so that the weld stands about one-eighth inch above the surrounding surface of the work.

Various forms of flux may be used and they are applied by dipping the end of the welding rod into the powder at intervals. These powders may contain borax or salt, and to prevent a hard, brittle weld, graphite or ferro-silicon may be added. Flux should be added only after the iron is molten and as little as possible should be used. No flux should be used just before completion of the work.

The welding flame should be played on the work around the crack and gradually brought to bear on the work. The bottom of the bevel should be joined first and it will be noted that the cast iron tends to run toward the flame, but does not stick together easily. A hard and porous weld should be carefully guarded against, as described above, and upon completion of the work the welded surface should be scraped with a file, while still red hot, in order to remove the surface scale.

Malleable Iron.—This material should be beveled in the same way that cast iron is handled, and preheating and slow cooling are equally desirable. The flame used is the same as for cast iron and so is the flux. The welding rod may be of cast iron, although better results are secured with Norway iron wire or else a mild steel wire wrapped with a coil of copper wire.

It will be understood that malleable iron turns to ordinary cast iron when melted and cooled. Welds in malleable iron are usually far from satisfactory and a better joint is secured by brazing the edges together with bronze. The edges to be joined are brought to a heat just a little below the point at which they will flow and the opening is then quickly-filled from a rod of Tobin bronze or manganese bronze, a brass or bronze flux being used in this work.

Wrought Iron or Semi-Steel.—This metal should be beveled and heated in the same way as described for cast iron. The flame should be neutral, of the same size as for steel, and used with the tip of the blue-white cone just touching the work. The welding rod should be of mild steel, or, if wrought iron is to be welded to steel, a cast iron rod may be used. A cast iron flux is well suited for this work. It should be noted that wrought iron turns to ordinary cast iron if kept heated for any length of time.

Steel.—Steel should be beveled if more than one-eighth inch in thickness. It requires only a local preheating around the point to be welded. The welding flame should be absolutely neutral, without excess of either gas. If the metal is one-sixteenth inch or less in thickness, the tip of the blue-white cone must be held a short distance from the surface of the work; in all other cases the tip of this cone is touched to the metal being welded.

The welding rod may be of mild, low carbon steel or of Norway iron. Nickel steel rods may be used for parts requiring great strength, but vanadium alloys are very difficult to handle. A very satisfactory rod is made by twisting together two wires of the required material. The rod must be kept constantly in contact with the work and should not be added until the edges are thoroughly melted. The flux may or may not be used. If one is wanted, it may be made from three parts iron filings, six parts borax and one part sal ammoniac.

It will be noticed that the steel runs from the flame, but tends to hold together. Should foaming commence in the molten metal, it shows an excess of oxygen and that the metal is being burned.

High carbon steels are very difficult to handle. It is claimed that a drop or two of copper added to the weld will assist the flow, but will also harden the work. An excess of oxygen reduces the amount of carbon and softens the steel, while an excess of acetylene increases the proportion of carbon and hardens the metal. High speed steels may sometimes be welded if first coated with semi-steel before welding.

Aluminum.—This is the most difficult of the commonly found metals to weld. This is caused by its high rate of expansion and contraction and its liability to melt and fall away from under the flame. The aluminum seems to melt on the inside first, and, without previous warning, a portion of the work will simply vanish from in front of the operator's eyes. The metal tends to run from the flame and separate at the same time. To keep the metal in shape and free from oxide, it is worked or puddled while in a plastic condition by an iron rod which has been flattened at one end. Several of these rods should be at hand and may be kept in a jar of salt water while not being used. These rods must not become coated with aluminum and they must not get red hot while in the weld.

The surfaces to be joined, together with the adjacent parts, should be cleaned thoroughly and then washed with a 25 per cent solution of nitric acid in hot water, used on a swab. The parts should then be rinsed in clean water and dried with sawdust. It is also well to make temporary fire clay moulds back of the parts to be heated, so that the metal may be flowed into place and allowed to cool without danger of breakage.

Aluminum must invariably be preheated to about 600 degrees, and the whole piece being handled should be well covered with sheet asbestos to prevent excessive heat radiation.

The flame is formed with an excess of acetylene such that the second cone extends about an inch, or slightly more, beyond the small blue-white point. The torch should be held so that the end of this second cone is in contact with the work, the small cone ordinarily used being kept an inch or an inch and a half from the surface of the work.

Welding rods of special aluminum are used and must be handled with their end submerged in the molten metal of the weld at all times.

When aluminum is melted it forms alumina, an oxide of the metal. This alumina surrounds small masses of the metal, and as it does not melt at temperatures below 5000 degrees (while aluminum melts at about 1200), it prevents a weld from being made. The formation of this oxide is retarded and the oxide itself is dissolved by a suitable flux, which usually contains phosphorus to break down the alumina.

Copper.—The whole piece should be preheated and kept well covered while welding. The flame must be much larger than for the same thickness of steel and neutral in character. A slight excess of acetylene would be preferable to an excess of oxygen, and in all cases the molten metal should be kept enveloped with the flame. The welding rod is of copper which contains phosphorus; and a flux, also containing phosphorus, should be spread for about an inch each side of the joint. These assist in preventing oxidation, which is sure to occur with heated copper.

Copper breaks very easily at a heat slightly under the welding temperature and after cooling it is simply cast copper in all cases.

Brass and Bronze.—It is necessary to preheat these metals, although not to a very high temperature. They must be kept well covered at all times to prevent undue radiation. The flame should be produced with a nozzle one size larger than for the same thickness of steel and the small blue-white cone should be held from one-fourth to one-half inch above the surface of the work. The flame should be neutral in character.

A rod or wire of soft brass containing a large percentage of zinc is suitable for adding to brass, while copper requires the use of copper or manganese bronze rods. Special flux or borax may be used to assist the flow.

The emission of white smoke indicates that the zinc contained in these alloys is being burned away and the heat should immediately be turned away or reduced. The fumes from brass and bronze welding are very poisonous and should not be breathed.

RESTORATION OF STEEL

The result of the high heat to which the steel has been subjected is that it is weakened and of a different character than before welding. The operator may avoid this as much as possible by first playing the outer flame of the torch all over the surfaces of the work just completed until these faces are all of uniform color, after which the metal should be well covered with asbestos and allowed to cool without being disturbed. If a temporary heating oven has been employed, the work and oven should be allowed to cool together while protected with the sheet asbestos. If the outside air strikes the freshly welded work, even for a moment, the result will be breakage.

A weld in steel will always leave the metal with a coarse grain and with all the characteristics of rather low grade cast steel. As previously mentioned in another chapter, the larger the grain size in steel the weaker the metal will be, and it is the purpose of the good workman to avoid, as far as possible, this weakening.

The structure of the metal in one piece of steel will differ according to the heat that it has under gone. The parts of the work that have been at the melting point will, therefore, have the largest grain size and the least strength. Those parts that have not suffered any great rise in temperature will be practically unaffected, and all the parts between these two extremes will be weaker or stronger according to their distance from the weld itself. To restore the steel so that it will have the best grain size, the operator may resort to either of two methods: (1) The grain may be improved by forging. That means that the metal added to the weld and the surfaces that have been at the welding heat are hammered much as a blacksmith would hammer his finished work to give it greater strength. The hammering should continue from the time the metal first starts to cool until it has reached the temperature at which the grain size is best for strength. This temperature will vary somewhat with the composition of the metal being handled, but in a general way, it may be stated that the hammering should continue without intermission from the time the flame is removed from the weld until the steel just begins to show attraction for a magnet presented to it. This temperature of magnetic attraction will always be low enough and the hammering should be immediately discontinued at this point. (2) A method that is more satisfactory, although harder to apply, is that of reheating the steel to a certain temperature throughout its whole mass where the heat has had any effect, and then allowing slow and even cooling from this temperature. The grain size is affected by the temperature at which the reheating is stopped, and not by the cooling, yet the cooling should be slow enough to avoid strains caused by uneven contraction.

After the weld has been completed the steel must be allowed to cool until below 1200° Fahrenheit. The next step is to heat the work slowly until all those parts to be restored have reached a temperature at which the magnet just ceases to be attracted. While the very best temperature will vary according to the nature and hardness of the steel being handled, it will be safe to carry the heating to the point indicated by the magnet in the absence of suitable means of measuring accurately these high temperatures. In using a magnet for testing, it will be most satisfactory if it is an electromagnet and not of the permanent type. The electric current may be secured from any small battery and will be the means of making sure of the test. The permanent magnet will quickly lose its power of attraction under the combined action of the heat and the jarring to which it will be subjected.

In reheating the work it is necessary to make sure that no part reaches a temperature above that desired for best grain size and also to see that all parts are brought to this temperature. Here enters the greatest difficulty in restoring the metal. The heating may be done so slowly that no part of the work on the outside reaches too high a temperature and then keeps the outside at this heat until the entire mass is at the same temperature. A less desirable way is to heat the outside higher than this temperature and allow the conductivity of the metal to distribute the excess to the inside.

The most satisfactory method, where it can be employed, is to make use of a bath of some molten metal or some chemical mixture that can be kept at the exact heat necessary by means of gas fires that admit of close regulation. The temperature of these baths may be maintained at a constant point by watching a pyrometer, and the finished work may be allowed to remain in the bath until all parts have reached the desired temperature.

WELDING INFORMATION

The following tables include much of the information that the operator must use continually to handle the various metals successfully. The temperature scales are given for convenience only. The composition of various alloys will give an idea of the difficulties to be contended with by consulting the information on welding various metals. The remaining tables are of self-evident value in this work.

TEMPERATURE SCALES

Centigrade Fahrenheit Centigrade Fahrenheit

200° 392° 1000° 1832°
225° 437° 1050° 1922°
250° 482° 1100° 2012°
275° 527° 1150° 2102°
300° 572° 1200° 2192°
325° 617° 1250° 2282°
350° 662° 1300° 2372°
375° 707° 1350° 2462°
400° 752° 1400° 2552°
425° 797° 1450° 2642°
450° 842° 1500° 2732°
475° 887° 1550° 2822°
500° 932° 1600° 2912°
525° 977° 1650° 3002°
550° 1022° 1700° 3092°
575° 1067° 1750° 3182°
600° 1112° 1800° 3272°
625° 1157° 1850° 3362°
650° 1202° 1900° 3452°
675° 1247° 2000° 3632°
700° 1292° 2050° 3722°
725° 1337° 2100° 3812°
750° 1382° 2150° 3902°
775° 1427° 2200° 3992°
800° 1472° 2250° 4082°
825° 1517° 2300° 4172°
850° 1562° 2350° 4262°
875° 1607° 2400° 4352°
900° 1652° 2450° 4442°
925° 1697° 2500° 4532°
950° 1742° 2550° 4622°
975° 1787° 2600° 4712°

METAL ALLOYS
(Society of Automobile Engineers)

Babbitt—
  Tin 84.00%
  Antimony 9.00%
  Copper 7.00%

Brass, White—
  Copper .... 3.00% to 6.00%
  Tin (minimum) 65.00%
  Zinc .... 28.00% to 30.00%

Brass, Red Cast—
  Copper 85.00%
  Tin 5.00%
  Lead 5.00%
  Zinc 5.00%

Brass, Yellow—
  Copper .... 62.00% to 65.00%
  Lead .... 2.00% to 4.00%
  Zinc .... 36.00% to 31.00%

Bronze, Hard—
  Copper .... 87.00% to 88.00%
  Tin .... 9.50% to 10.50%
  Zinc .... 1.50% to 2.50%

Bronze, Phosphor—
  Copper 80.00%
  Tin 10.00%
  Lead 10.00%
  Phosphorus .... .50% to .25%

Bronze, Manganese—
  Copper (approximate) 60.00%
  Zinc (approximate) 40.00%

Manganese (variable) .... small

Bronze, Gear —
  Copper......................... 88.00% to 89.00%
  Tin............................ 11.00% to 12.00%

Aluminum Alloys —
    Aluminum Copper Zinc Manganese
No. 1.. 90.00% 8.5-7.0%
No. 2.. 80.00% 2.0-3.0% 15% Not over 0.40%
No. 3.. 65.00% 35.0%

Cast Iron —
    Gray Iron Malleable
Total carbon........3.0 to 3.5%
Combined carbon.....0.4 to 0.7%
Manganese...........0.4 to 0.7% 0.3 to 0.7%
Phosphorus..........0.6 to 1.0% Not over 0.2%
Sulphur...........Not over 0.1% Not over 0.6%
Silicon............1.75 to 2.25% Not over 1.0%

Carbon Steel (10 Point) —
  Carbon......................... .05% to .15%
  Manganese...................... .30% to .60%
  Phosphorus (maximum).......... .045%
  Sulphur (maximum)............. .05%
(20 Point) —
  Carbon......................... .15% to .25%
  Manganese...................... .30% to .60%
  Phosphorus (maximum).......... .045%
  Sulphur (maximum)............. .05%
(35 Point) —
  Manganese...................... .50% to .80%
  Carbon......................... .30% to .40%
  Phosphorus (maximum).......... .05%
  Sulphur (maximum)............. .05%
(95 Point) —
  Carbon......................... .90% to 1.05%
  Manganese...................... .25% to .50%

Phosphorus (maximum)........... .04%
Sulphur (maximum)............. .05%

HEATING POWER OF FUEL GASES

(In B.T.U. per Cubic Foot.)
Acetylene....... 1498.99 Ethylene....... 1562.9
Hydrogen........ 291.96 Methane........ 953.6
Alcohol......... 1501.76

MELTING POINTS OF METALS
Platinum....................3200°
Iron, wrought...............2900°
  malleable.................2500°
  cast......................2400°
  pure......................2760°
Steel, mild.................2700°
  Medium....................2600°
  Hard......................2500°
Copper......................1950°
Brass.......................1800°
Silver......................1750°
Bronze......................1700°
Aluminum....................1175°
Antimony....................1150°
Zinc......................... 800°
Lead......................... 620°
Babbitt.................500-700°
Solder..................500-575°
Tin.......................... 450°

NOTE. – These melting points are for average compositions and conditions. The exact proportion of elements entering into the metals affects their melting points one way or the other in practice.

TENSILE STRENGTH OF METALS

Alloy steels can be made with tensile strengths as high as 300,000 pounds per square inch. Some carbon steels are given below according to "points":

Pounds per Square Inch

Steel, 10 point	50,000 to 65,000
20 point	60,000 to 80,000
40 point	70,000 to 100,000
60 point	90,000 to 120,000
Iron, Cast	13,000 to 30,000
Wrought	40,000 to 60,000
Malleable	25,000 to 45,000
Copper	24,000 to 50,000
Bronze	30,000 to 60,000
Brass, Cast	12,000 to 18,000
Rolled	30,000 to 40,000
Wire	60,000 to 75,000
Aluminum	12,000 to 23,000
Zinc	5,000 to 15,000
Tin	3,000 to 5,000
Lead	1,500 to 2,500

CONDUCTIVITY OF METALS

(Based on the Value of Silver as 100)

	Heat	Electricity
Silver	100	100
Copper	74	99
Aluminum	38	63
Brass	23	22
Zinc	19	29
Tin	14	15
Wrought Iron	12	16
Steel	11.5	12
Cast Iron	11	12
Bronze	9	7

Lead...................... 8 9

WEIGHT OF METALS

(Per Cubic Inch)
　　　　Pounds　Pounds
Lead............ .410 Wrought Iron..... .278
Copper.......... .320 Tin.............. .263
Bronze.......... .313 Cast Iron........ .260
Brass........... .300 Zinc............. .258
Steel........... .283 Aluminum......... .093

EXPANSION OF METALS

(Measured in Thousandths of an Inch per Foot of Length When Raised 1000 Degrees in Temperature)
　　　　　Inch　Inch
Lead............ .188 Brass............ .115
Zinc............ .168 Copper........... .106
Aluminum........ .148 Steel............ .083
Silver.......... .129 Wrought Iron..... .078
Bronze.......... .118 Cast Iron........ .068

CHAPTER VI

ELECTRIC WELDING

RESISTANCE METHOD

Two distinct forms of electric welding apparatus are in use, one producing heat by the resistance of the metal being treated to the passage of electric current, the other using the heat of the electric arc.

The resistance process is of the greatest use in manufacturing lines where there is a large quantity of one kind of work to do, many thousand pieces of one kind, for instance. The arc method may be applied in practically any case where any other form of weld may be made. The resistance process will be described first.

It is a well known fact that a poor conductor of electricity will offer so much resistance to the flow of electricity that it will heat. Copper is a good conductor, and a bar of iron, a comparatively poor conductor, when placed between heavy copper conductors of a welder, becomes heated in attempting to carry the large volume of current. The degree of heat depends on the amount of current and the resistance of the conductor.

In an electric circuit the ends of two pieces of metal brought together form the point of greatest resistance in the electric circuit, and the abutting ends instantly begin to heat. The hotter this metal becomes, the greater the resistance to the flow of current; consequently, as the edges of the abutting ends heat, the current is forced into the adjacent cooler parts, until there is a uniform heat throughout the entire mass. The heat is first developed in the interior of the metal so that it is welded there as perfectly as at the surface.

[Illustration: Figure 42.—Spot Welding Machine]

The electric welder (Figure 42) is built to hold the parts to be joined between two heavy copper dies or contacts. A current of three to five volts, but of very great volume (amperage), is allowed to pass across these dies, and in going through the metal to be welded, heats the edges to a welding temperature. It may be explained that the voltage of an electric current measures the pressure or force with which it is being sent through the circuit and has nothing to do with the quantity or volume passing. Amperes measure the rate at which the current is passing through the circuit and consequently give a measure of the quantity which passes in any given time. Volts correspond to water pressure measured by pounds to the square inch; amperes represent the flow in gallons per minute. The low voltage used avoids all danger to the operator, this pressure not being sufficient to be felt even with the hands resting on the copper contacts.

Current is supplied to the welding machine at a higher voltage and lower amperage than is actually used between the dies, the low voltage and high amperage being produced by a transformer incorporated in the machine itself. By means of windings of suitable size wire, the outside current may be received at voltages ranging from 110 to 550 and converted to the low pressure needed.

The source of current for the resistance welder must be alternating, that is, the current must first be negative in value and then positive, passing from one extreme to the other at rates varying from 25 to 133 times a second. This form is known as alternating current, as opposed to direct current, in which there is no changing of positive and negative.

The current must also be what is known as single phase, that is, a current which rises from zero in value to the highest point as a positive current and then recedes to zero before rising to the highest point of negative value. Two-phase of three-phase currents would give two or three positive impulses during this time.

As long as the current is single phase alternating, the voltage and cycles (number of alternations per second) may be anything convenient. Various voltages and cycles are taken care of by specifying all these points when designing the transformer which is to handle the current.

Direct current is not used because there is no way of reducing the voltage conveniently without placing resistance wires in the circuit and this uses power without producing useful work. Direct current may be changed to alternating by having a direct current motor running an alternating current dynamo, or the change may be made by a rotary converter, although this last method is not so satisfactory as the first.

The voltage used in welding being so low to start with, it is absolutely necessary that it be maintained at the correct point. If the source of current supply is not of ample capacity for the welder being used, it will be very hard to avoid a fall of voltage when the current is forced to pass through the high resistance of the weld. The current voltage for various work is calculated accurately, and the efficiency of the outfit depends to a great extent on the voltage being constant.

A simple test for fall of voltage is made by connecting an incandescent electric lamp across the supply lines at some point near the welder. The lamp should burn with the same brilliancy when the weld is being made as at any other time. If the lamp burns dim at any time, it indicates a drop in voltage, and this condition should be corrected.

The dynamo furnishing the alternating current may be in the same building with the welder and operated from a direct current motor, as mentioned above, or operated from any convenient shafting or source of power. When the dynamo is a part of the welding plant it should be placed as close to the welding machine as possible, because the length of the wire used affects the voltage appreciably.

In order to hold the voltage constant, the Toledo Electric Welder Company has devised connections which include a rheostat to insert a variable resistance in the field windings of the dynamo so that the voltage may be increased by cutting this resistance out at the proper time. An auxiliary switch is connected to the welder switch so that both switches act together. When the welder switch is closed in making a weld, that portion of the rheostat resistance between two arms determining the voltage is short circuited. This lowers the resistance and the field magnets of the dynamo are made stronger so that additional voltage is provided to care for the resistance in the metal being heated.

A typical machine is shown in the accompanying cut (Figure 43). On top of the welder are two jaws for holding the ends of the pieces to be welded. The lower part of the jaws is rigid while the top is brought down on top of the work, acting as a clamp. These jaws carry the copper dies through which the current enters the work being handled. After the work is clamped between the jaws, the upper set is forced closer to the lower set by a long compression lever. The current being turned on with the surfaces of the work in contact, they immediately heat to the welding point when added pressure on the lever forces them together and completes the weld.

[Illustration: Figure 43 — Operating Parts of a Toledo Spot Welder]

[Illustration: Figure 43a. — Method of Testing Electric Welder]
[Illustration: Figure 44. — Detail of Water-Cooled Spot Welding Head]

The transformer is carried in the base of the machine and on the left-hand side is a regulator for controlling the voltage for various kinds of work. The clamps are applied by treadles convenient to the foot of the operator. A treadle is provided which instantly releases both jaws upon the completion of the weld. One or both of the copper dies may be cooled by a stream of water circulating through it from the city water mains (Figure 44). The regulator and switch give the operator control of the heat, anything from a dull red to the melting point being easily obtained by movement of the lever (figure 45).

[Illustration: Figure 45. — Welding Head of a Water-Cooled Welder]

Welding. — It is not necessary to give the metal to be welded any special preparation, although when very rusty or covered with scale, the rust and scale should be removed sufficiently to allow good contact of clean metal on the copper dies. The cleaner and better the stock, the less current it takes, and there is less wear on the dies. The dies should be kept firm and tight in their holders to make a good contact. All bolts and nuts fastening the electrical contacts should be clean and tight at all times.

The scale may be removed from forgings by immersing them in a pickling solution in a wood, stone or lead-lined tank.

The solution is made with five gallons of commercial sulphuric acid in 150 gallons of water. To get the quickest and best results from this method, the solution should be kept as near the boiling point as possible by having a coil of extra heavy lead pipe running inside the tank and carrying live steam. A very few minutes in this bath will remove the scale and the parts should then be washed in running water. After this washing they should be dipped into a bath of 50 pounds of unslaked lime in 150 gallons of water to neutralize any trace of acid.

Cast iron cannot be commercially welded, as it is high in carbon and silicon, and passes suddenly from a crystalline to a fluid state when brought to the welding temperature. With steel or wrought iron the temperature must be kept below the melting point to avoid injury to the metal. The metal must be heated quickly and pressed together with sufficient force to push all burnt metal out of the joint.

High carbon steel can be welded, but must be annealed after welding to overcome the strains set up by the heat being applied at one place. Good results are hard to obtain when the carbon runs as high as 75 points, and steel of this class can only be handled by an experienced operator. If the steel is below 25 points in carbon content, good welds will always be the result. To weld high carbon to low carbon steel, the stock should be clamped in the dies with the low carbon stock sticking considerably further out from the die than the high carbon stock. Nickel steel welds readily, the nickel increasing the strength of the weld.

Iron and copper may be welded together by reducing the size of the copper end where it comes in contact with the iron. When welding copper and brass the pressure must be less than when welding iron. The metal is allowed to actually fuse or melt at the juncture and the pressure must be sufficient to force the burned metal out. The current is cut off the instant the metal ends begin to soften, this being done by means of an automatic switch which opens when the softening of the metal allows the ends to come together. The pressure is applied to the weld by having the sliding jaw moved by a weight on the end of an arm.

Copper and brass require a larger volume of current at a lower voltage than for steel and iron. The die faces are set apart three times the diameter of the stock for brass and four times the diameter for copper.

Light gauges of sheet steel can be welded to heavy gauges or to solid bars of steel by "spot" welding, which will be described later. Galvanized iron can be welded, but the zinc coating will be burned off. Sheet steel can be welded to cast iron, but will pull apart, tearing out particles of the iron.

Sheet copper and sheet brass may be welded, although this work requires more experience than with iron and steel. Some grades of sheet aluminum can be spot-welded if the slight roughness left on the surface under the die is not objectionable.

Butt Welding.—This is the process which joins the ends of two pieces of metal as described in the foregoing part of this chapter. The ends are in plain sight of the operator at all times and it can easily be seen when the metal reaches the welding heat and begins to soften (Figure 46). It is at this point that the pressure must be applied with the lever and the ends forced together in the weld.

[Illustration: Figure 46.—Butt Welder]

The parts are placed in the clamping jaws (Figure 47) with 1/8 to 1/2 inch of metal extending beyond the jaw. The ends of the metal touch each other and the current is turned on by means of a switch. To raise the ends to the proper heat requires from 3 seconds for 1/4-inch rods to 35 seconds for a 1-1/2-inch bar.

This method is applicable to metals having practically the same area of metal to be brought into contact on each end. When such parts are forced together a slight projection will be left in the form of a fin or an enlarged portion called an upset. The degree of heat required for any work is found by moving the handle of the regulator one way or the other while testing several parts. When this setting is right the work can continue as long as the same sizes are being handled.

[Illustration: Figure 47.—Clamping Dies of a Butt Welder]

Copper, brass, tool steel and all other metals that are harmed by high temperatures must be heated quickly and pressed together with sufficient force to force all burned metal from the weld.

In case it is desired to make a weld in the form of a capital letter T, it is necessary to heat the part corresponding to the top bar of the T to a bright red, then bring the lower bar to the pre-heated one and again turn on the current, when a weld can be quickly made.

Spot Welding.—This is a method of joining metal sheets together at any desired point by a welded spot about the size of a rivet. It is done on a spot welder by fusing the metal at the point desired and at the same instant applying sufficient pressure to force the particles of molten metal together. The dies are usually placed one above the other so that the work may rest on the lower one while the upper one is brought down on top of the upper sheet to be welded.

One of the dies is usually pointed slightly, the opposing one being left flat. The pointed die leaves a slight indentation on one side of the metal, while the other side is left smooth. The dies may be reversed so that the outside surface of any work may be left smooth. The current is allowed to flow through the dies by a switch which is closed after pressure is applied to the work.

There is a limit to the thickness of sheet metal that can be welded by this process because of the fact that the copper rods can only carry a certain quantity of current without becoming unduly heated themselves. Another reason is that it is difficult to make heavy sections of metal touch at the welding point without excessive pressure.

Lap welding is the process used when two pieces of metal are caused to overlap and when brought to a welding heat are forced together by passing through rollers, or under a press, thus leaving the welded joint practically the same thickness as the balance of the work.

Where it is desirable to make a continuous seam, a special machine is required, or an attachment for one of the other types. In this form of work the stock must be thoroughly cleaned and is then passed between copper rollers which act in the same capacity as the copper dies.

Other Applications.—Hardening and tempering can be done by clamping the work in the welding dies and setting the control and time to bring the metal to the proper color, when it is cooled in the usual manner.

Brazing is done by clamping the work in the jaws and heating until the flux, then the spelter has melted and run into the joint. Riveting and heading of rivets can be done by bringing the dies down on opposite ends of the rivet after it has been inserted in the hole, the dies being shaped to form the heads properly.

Hardened steel may be softened and annealed so that it can be machined by connecting the dies of the welder to each side of the point to be softened. The current is then applied until the work has reached a point at which it will soften when cooled.

Troubles and Remedies.—The following methods have been furnished by the Toledo Electric Welder Company and are recommended for this class of work whenever necessary.

To locate grounds in the primary or high voltage side of the circuit, connect incandescent lamps in series by means of a long piece of lamp cord, as shown, in Figure 43a. For 110 volts use one lamp, for 220 volts use two lamps and for 440 volts use four lamps. Attach one end of the lamp cord to one side of the switch, and close the switch. Take the other end of the cord in the hand and press it against some part of the welder frame where the metal is clean and bright. Paint, grease and dirt act as insulators and prevent electrical contact. If the lamp lights, the circuit is in electrical contact with the frame; in other words, grounded. If the lamps do not light, connect the wire to a terminal block, die or slide. If the lamps then light, the circuit, coils or leads are in electrical contact with the large coil in the transformer or its connections.

If, however, the lamps do not light in either case, the lamp cord should be disconnected from the switch and connected to the other side, and the operations of connecting to welder frame, dies, terminal blocks, etc., as explained above, should be repeated. If the lamps light at any of these connections, a "ground" is indicated. "Grounds" can usually be found by carefully tracing the primary circuit until a place is found where the insulation is defective. Reinsulate and make the above tests again to make sure everything is clear. If the ground can not be located by observation, the various parts of the primary circuit should be disconnected, and the transformer, switch, regulator, etc., tested separately.

To locate a ground in the regulator or other part, disconnect the lines running to the welder from the switch. The test lamps used in the previous tests are connected, one end of lamp cord to the switch, the other end to a binding post of the regulator. Connect the other side of the switch to some part of the regulator housing. (This must be a clean connection to a bolt head or the paint should be scraped off.) Close the switch. If the lamps light, the regulator winding or some part of the switch is "grounded" to the iron base or core of the regulator. If the lamps do not light, this part of the apparatus is clear.

This test can be easily applied to any part of the welder outfit by connecting to the current carrying part of the apparatus, and to the iron base or frame that should not carry current. If the lamps light, it indicates that the insulation is broken down or is defective.

An A.C. voltmeter can, of course, be substituted for the lamps, or a D.C. voltmeter with D.C. current can be used in making the tests.

A short circuit in the primary is caused by the insulation of the coils becoming defective and allowing the bare copper wires to touch each other. This may result in a "burn out" of one or more of the transformer coils, if the trouble is in the transformer, or in the continued blowing of fuses in the line. Feel of each coil separately. If a short circuit exists in a coil it will heat excessively. Examine all the wires; the insulation may have worn through and two of them may cross, or be in contact with the frame or other part of the welder. A short circuit in the regulator winding is indicated by failure of the apparatus to regulate properly, and sometimes, though not always, by the heating of the regulator coils.

The remedy for a short circuit is to reinsulate the defective parts. It is a good plan to prevent trouble by examining the wiring occasionally and see that the insulation is perfect.

To Locate Grounds and Short Circuits in the Secondary, or Low Voltage Side.—Trouble of this kind is indicated by the machine acting sluggish or, perhaps, refusing to operate. To make a test, it will be necessary to first ascertain the exciting current of your particular transformer. This is the current the transformer draws on "open circuit," or when supplied with current from the line with no stock in the welder dies. The following table will give this information close enough for all practical purposes:

K.W. Rating	Amperes at 110 Volts	220 Volts	440 Volts	550 Volts
3	1.5	.75	.38	.3
5	2.5	1.25	.63	.5
8	3.6	1.8	.9	.72
10	4.25	2.13	1.07	.85
15	6.	3.	1.5	1.2
20	7.	3.5	1.75	1.4
30	9.	4.5	2.25	1.8
35	9.6	4.8	2.4	1.92
50	10.	5.	2.5	2

Remove the fuses from the wall switch and substitute fuses just large enough to carry the "exciting" current. If no suitable fuses are at hand, fine strands of copper from an ordinary lamp cord may be used. These strands are usually No. 30 gauge wire and will fuse at about 10 amperes. One or more strands should be used, depending on the amount of exciting current, and are connected across the fuse clips in place of fuse wire. Place a piece of wood or fibre between the welding dies in the welder as though you were going to weld them. See that the regulator is on the highest point and close the welder switch. If the secondary circuit is badly grounded, current will flow through the ground, and the small fuses or small strands of wire will burn out. This is an indication that both sides of the secondary circuit are grounded or that a short circuit exists in a primary coil. In either case the welder should not be operated until the trouble is found and removed. If, however, the small fuses do not "blow," remove same and replace the large fuses, then disconnect wires running from the wall switch to the welder and substitute two pieces of No. 8 or No. 6 insulated copper wire, after scraping off the insulation for an inch or two at each end. Connect one wire from the switch to the frame of welder; this will leave one loose end. Hold this a foot or so away from the place where the insulation is cut off; then turn on the current and strike the free end of this wire lightly against one of the copper dies, drawing it away quickly. If no sparking is produced, the secondary circuit is free from ground, and you will then look for a broken connection in the circuit. Some caution must be used in making the above test, as in case one terminal is heavily grounded the testing wire may be fused if allowed to stay in contact with the die.

The Remedy.—Clean the slides, dies and terminal blocks thoroughly and dry out the fibre insulation if it is damp. See that no scale or metal has worked under the sliding parts, and that the secondary leads do not touch the frame. If the ground is very heavy it may be necessary to remove the slides in order to facilitate the examination and removal of the ground. Insulation, where torn or worn through, must be carefully replaced or taped. If the transformer coils are grounded to the iron core of the transformer or to the secondary, it may be necessary to remove the coils and reinsulate them at the points of contact. A short circuited coil will heat excessively and eventually burn out. This may mean a new coil if you are unable to repair the old one. In all cases the transformer windings should be protected from mechanical injury or dampness. Unless excessively overloaded, transformers will last for years without giving a moment's trouble, if they are not exposed to moisture or are not injured mechanically.

The most common trouble arises from poor electrical contacts, and they are the cause of endless trouble and annoyance. See that all connections are clean and bright. Take out the dies every day or two and see that there is no scale, grease or dirt between them and the holders. Clean them thoroughly before replacing. Tighten the bolts running from the transformer leads to the work jaws.

ELECTRIC ARC WELDING

This method bears no relation to the one just considered, except that the source of heat is the same in both cases. Arc welding makes use of the flame produced by the voltaic arc in practically the same way that oxy-acetylene welding uses the flame from the gases.

If the ends of two pieces of carbon through which a current of electricity is flowing while they are in contact are separated from each other quite slowly, a brilliant arc of flame is formed between them which consists mainly of carbon vapor. The carbons are consumed by combination with the oxygen in the air and through being turned to a gas under the intense heat.

The most intense action takes place at the center of the carbon which carries the positive current and this is the point of greatest heat. The temperature at this point in the arc is greater than can be produced by any other means under human control.

An arc may be formed between pieces of metal, called electrodes, in the same way as between carbon. The metallic arc is called a flaming arc and as the metal of the electrode burns with the heat, it gives the flame a color characteristic of the material being used. The metallic arc may be drawn out to a much greater length than one formed between carbon electrodes.

Arc Welding is carried out by drawing a piece of carbon which is of negative polarity away from the pieces of metal to be welded while the metal is made positive in polarity. The negative wire is fastened to the carbon electrode and the work is laid on a table made of cast or wrought iron to which the positive wire is made fast. The direction of the flame is then from the metal being welded to the carbon and the work is thus prevented from being saturated with carbon, which would prove very detrimental to its strength. A secondary advantage is found in the fact that the greatest heat is at the metal being welded because of its being the positive electrode.

The carbon electrode is usually made from one quarter to one and a half inches in diameter and from six to twelve inches in length. The length of the arc may be anywhere from one inch to four inches, depending on the size of the work being handled.

While the parts are carefully insulated to avoid danger of shock, it is necessary for the operator to wear rubber gloves as a further protection, and to wear some form of hood over the head to shield him against the extreme heat liberated. This hood may be made from metal, although some material that does not conduct electricity is to be preferred. The work is watched through pieces of glass formed with one sheet, which is either blue or green, placed over another which is red. Screens of glass are sometimes used without the head protector. Some protection for the eyes is absolutely necessary because of the intense white light.

It is seldom necessary to preheat the work as with the gas processes, because the heat is localized at the point of welding and the action is so rapid that the expansion is not so great. The necessity of preheating, however, depends entirely on the material, form and size of the work being handled. The same advice applies to arc welding as to the gas flame method but in a lesser degree. Filling rods are used in the same way as with any other flame process.

It is the purpose of this explanation to state the fundamental principles of the application of the electric arc to welding metals, and by applying the principles the following questions will be answered:

What metals can be welded by the electric arc?

What difficulties are to be encountered in applying the electric arc to welding?

What is the strength of the weld in comparison with the original piece?

What is the function of the arc welding machine itself?

What is the comparative application of the electric arc and the oxy-acetylene method and others of a similar nature?

The answers to these questions will make it possible to understand the application of this process to any work. In a great many places the use of the arc is cutting the cost of welding to a very small fraction of what it would be by any other method, so that the importance of this method may be well understood.

Any two metals which are brought to the melting temperature and applied to each other will adhere so that they are no more apt to break at the weld than at any other point outside of the weld. It is the property of all metals to stick together under these conditions. The electric arc is used in this connection merely as a heating agent. This is its only function in the process.

It has advantages in its ease of application and the cheapness with which heat can be liberated at any given point by its use. There is nothing in connection with arc welding that the above principles will not answer; that is, that metals at the melting point will weld and that the electric arc will furnish the heat to bring them to this point. As to the first question, what metals can be welded, all metals can be welded.

The difficulties which are encountered are as follows:

In the case of brass or zinc, the metals will be covered with a coat of zinc oxide before they reach a welding heat. This zinc oxide makes it impossible for two clean surfaces to come together and some method has to be used for eliminating this possibility and allowing the two surfaces to join without the possibility of the oxide intervening. The same is true of aluminum, in which the oxide, alumina, will be formed, and with several other alloys comprising elements of different melting points.

In order to eliminate these oxides, it is necessary in practical work, to puddle the weld; this is, to have a sufficient quantity of molten metal at the weld so that the oxide is floated away. When this is done, the two surfaces which are to be joined are covered with a coat of melted metal on which floats the oxide and other impurities. The two pieces are thus allowed to join while their surfaces are protected. This precaution is not necessary in working with steel except in extreme cases.

Another difficulty which is met with in the welding of a great many metals is their expansion under heat, which results in so great a contraction when the weld cools that the metal is left with a considerable strain on it. In extreme cases this will result in cracking at the weld or near it. To eliminate this danger it is necessary to apply heat either all over the piece to be welded or at certain points. In the case of cast iron and sometimes with copper it is necessary to anneal after welding, since otherwise the welded pieces will be very brittle on account of the chilling. This is also true of malleable iron.

Very thin metals which are welded together and are not backed up by something to carry away the excess heat, are very apt to burn through, leaving a hole where the weld should be. This difficulty can be eliminated by backing up the weld with a metal face or by decreasing the intensity of the arc so that this melting through will not occur. However, the practical limit for arc welding without backing up the work with a metal face or decreasing the intensity of the arc is approximately 22 gauge, although thinner metal can be welded by a very skillful and careful operator.

One difficulty with arc welding is the lack of skillful operators. This method is often looked upon as being something out of the ordinary and governed by laws entirely different from other welding. As a matter of fact, it does not take as much skill to make a good arc weld as it does to make a good weld in a forge fire as the blacksmith does it. There are few jobs which cannot be handled successfully by an operator of average intelligence with one week's instructions, although his work will become better and better in quality as he continues to use the arc.

Now comes the question of the strength of the weld after it has been made. This strength is equally as great as that of the metal that is used to make the weld. It should be remembered, however, that the metal which goes into the weld is put in there as a casting and has not been rolled. This would make the strength of the weld as great as the same metal that is used for filling if in the cast form.

Two pieces of steel could be welded together having a tensile strength at the weld of 50,000 pounds. Higher strengths than this can be obtained by the use of special alloys for the filling material or by rolling. Welds with a tensile strength as great as mentioned will give a result which is perfectly satisfactory in almost all cases.

There are a great many jobs where it is possible to fill up the weld, that is, make the section at the point of the weld a little larger than the section through the rest of the piece. By doing this, the disadvantages of the weld being in the form of a casting in comparison with the rest of the piece being in the form of rolled steel can be overcome, and make the weld itself even stronger than the original piece.

The next question is the adaptability of the electric arc in comparison with forge fire, oxy-acetylene or other method. The answer is somewhat difficult if made general. There are no doubt some cases where the use of a drop hammer and forge fire or the use of the oxy-acetylene torch will make, all things being considered, a better job than the use of the electric arc, although a case where this is absolutely proved is rare.

The electric arc will melt metal in a weld for less than the same metal can be melted by the use of the oxy-acetylene torch, and, on account of the fact that the heat can be applied exactly where it is required and in the amount required, the arc can in almost all cases supply welding heat for less cost than a forge fire or heating furnace.

The one great advantage of the oxy-acetylene method in comparison with other methods of welding is the fact that in some cases of very thin sheet, the weld can be made somewhat sooner than is possible otherwise. With metal of 18 gauge or thicker, this advantage is eliminated. In cutting steel, the oxy-acetylene torch is superior to almost any other possible method.

Arc Welding Machines. — A consideration of the function and purpose of the various types of arc welding machines shows that the only reason for the use of any machine is either for conversion of the current from alternating to direct, or, if the current is already direct, then the saving in the application of this current in the arc.

It is practically out of the question to apply an alternating current arc to welding for the reason that in any arc practically all the heat is liberated at the positive electrode, which means that, in alternating current, half the heat is liberated at each electrode as the current changes its direction of flow or alternates. Another disadvantage of the alternating arc is that it is difficult of control and application.

In all arc welding by the use of the carbon arc, the positive electrode is made the piece to be welded, while in welding with metallic electrodes this may be either the piece to be welded of the rod that is used as a filler. The voltage across the arc is a variable quantity, depending on the length of the flame, its temperature and the gases liberated in the arc. With a carbon electrode the voltage will vary from zero to forty-five volts. With the metallic electrode the voltage will vary from zero to thirty volts. It is, therefore, necessary for the welding machine to be able to furnish to the arc the requisite amount of current, this amount being varied, and furnish it at all times at the voltage required.

The simplest welding apparatus is a resistance in series with the arc. This is entirely satisfactory in every way except in cost of current. By the use of resistance in series with the arc and using 220 volts as the supply, from eighty to ninety per cent of the current is lost in heat at the resistance. Another disadvantage is the fact that most materials change their resistance as their temperature changes, thus making the amount of current for the arc a variable quantity, depending on the temperature of the resistance.

There have been various methods originated for saving the power mentioned and a good many machines have been put on the market for this purpose. All of them save some power over what a plain resistance would use. Practically all arc welding machines at the present time are motor generator sets, the motor of which is arranged for the supply voltage and current, this motor being direct connected to a compound wound generator delivering approximately seventy-five volts direct current. Then by the use of a resistance, this seventy-five volt supply is applied to the arc. Since the voltage across the arc will vary from zero to fifty volts, this machine will save from zero up to seventy per cent of the power that the machine delivers. The rest of the power, of course, has to be dissipated in the resistance used in series with the arc.

A motor generator set which can be purchased from any electrical company, with a long piece of fence wire wound around a piece of asbestos, gives results equally as good and at a very small part of the first cost.

It is possible to construct a machine which will eliminate all losses in the resistance; in other words, eliminate all resistance in series with the arc. A machine of this kind will save its cost within a very short time, providing the welder is used to any extent.

Putting it in figures, the results are as follows for average conditions. Current at 2c per kilowatt hour, metallic electrode arc of 150 amperes, carbon arc 500 amperes; voltage across the metallic electrode arc 20, voltage across the carbon arc 35. Supply current 220 volts, direct. In the case of the metallic electrode, if resistance is used, the cost of running this arc is sixty-six cents per hour. With the carbon electrode, $2.20 per hour. If a motor generator set with a seventy volt constant potential machine is used for a welder, the cost will be as follows:

Metallic electrode 25.2c. Carbon electrode 84c per hour. With a machine which will deliver the required voltage at the arc and eliminate all the resistance in series with the arc, the cost will be as follows: Metallic electrode 7.2c per hour; carbon electrode 42c per hour. This is with the understanding that the arc is held constant and continuously at its full value. This, however, is practically impossible and the actual load factor is approximately fifty per cent, which would mean that operating a welder as it is usually operated, this result will be reduced to one-half of that stated in all cases.

CHAPTER VII

HAND FORGING AND WELDING

Smithing, or blacksmithing, is the process of working heated iron, steel or other metals by forging, bending or welding them.

The Forge.—The metal is heated in a forge consisting of a shallow pan for holding the fire, in the center of which is an opening from below through which air is forced to make a hot fire.

[Illustration: Figure 48.—Tuyere Construction on a Forge]

Air is forced through this hole, called a "tuyere" (Figure 48) by means of a hand bellows, a rotary fan operated with crank or lever, or with a fan driven from an electric motor. The harder the air is driven into the fire above the tuyere the more oxygen is furnished and the hotter the fire becomes.

Directly below the tuyere is an opening through which the ashes that drop from the fire may be cleaned out.

The Fire.—The fire is made by placing a small piece of waste soaked in oil, kerosene or gasoline, over the tuyere, lighting the waste, then starting the fan or blower slowly. Gradually cover the waste, while it is burning brightly, with a layer of soft coal. The coal will catch fire and burn after the waste has been consumed. A piece of waste half the size of a person's hand is ample for this purpose.

The fuel should be "smithing coal." A lump of smithing coal breaks easily, shows clean and even on all sides and should not break into layers. The coal is broken into fine pieces and wet before being used on the fire.

The fire should be kept deep enough so that there is always three or four inches of fire below the piece of metal to be heated and there should be enough fire above the work so that no part of the metal being heated comes in contact with the air. The fire should be kept as small as possible while following these rules as to depth.

To make the fire larger, loosen the coal around the edges. To make the fire smaller, pack wet coal around the edges in a compact mass and loosen the fire in the center. Add fresh coal only around the edges of the fire. It will turn to coke and can then be raked onto the fire. Blow only enough air into the fire to keep it burning brightly, not so much that the fire is blown up through the top of the coal pack. To prevent the fire from going out between jobs, stick a piece of soft wood into it and cover with fresh wet coal.

Tools. — The *hammer* is a ball pene, or blacksmith's hammer, weighing about a pound and a half.

The *sledge* is a heavy hammer, weighing from 5 to 20 pounds and having a handle 30 to 36 inches long.

The *anvil* is a heavy piece of wrought iron (Figure 49), faced with steel and having four legs. It has a pointed horn on one end, an overhanging tail on the other end and a flat top. In the tail there is a square hole called the "hardie" hole and a round one called the "spud" hole.

[Illustration: Figure 49. — Anvil, Showing Horn, Tail, Hardie Hole and Spud Hole]

Tongs, with handles about one foot long and jaws suitable for holding the work, are used. To secure a firm grip on the work, the jaws may be heated red hot and hammered into shape over the piece to be held, thus giving a properly formed jaw. Jaws should touch the work along their entire length.

The *set hammer* is a hammer, one end of whose head is square and flat, and from this face the head tapers evenly to the other face. The large face is about 1-1/4 inches square.

The *flatter* is a hammer having one face of its head flat and about 2-1/2 inches square.

Swages are hammers having specially formed faces for finishing rounds, squares, hexagons, ovals, tapers, etc.

Fullers are hammers having a rounded face, long in one direction. They are used for spreading metal in one direction only.

The *hardy* is a form of chisel with a short, square shank which may be set into the hardie hole for cutting off hot bars.

Operations.—Blacksmithing consists of bending, drawing or upsetting with the various hammers, or in punching holes.

Bending is done over the square corners of the anvil if square cornered bends are desired, or over the horn of the anvil if rounding bends, eyes, hooks, etc., are wanted.

To bend a ring or eye in the end of a bar, first figure the length of stock needed by multiplying the diameter of the hole by 31/7, then heat the piece to a good full red at a point this distance back from the end. Next bend the iron over at a 90 degree angle (square) at this point. Next, heat the iron from the bend just made clear to the point and make the eye by laying the part that was bent square over the horn of the anvil and bending the extreme tip into part of a circle. Keep pushing the piece farther and farther over the horn of the anvil, bending it as you go. Do not hammer directly over the horn of the anvil, but on the side where you are doing the bending.

To make the outside of a bend square, sharp and full, rather than slightly rounding, the bent piece must be laid edgewise on the face of the anvil. That is, after making the bend over the corner of the anvil, lay the piece on top of the anvil so that its edge and not the flat side rests on the anvil top. With the work in this position, strike directly against the corner with the hammer so that the blows come in line, first with one leg of the work, then the other, and always directly on the corner of the piece. This operation cannot be performed by laying the work so that one leg hangs over the anvil's corner.

To make a shoulder on a rod or bar, heat the work and lay flat across the top of the anvil with the point at which the shoulder is desired at the edge of the anvil. Then place the set hammer on top of the piece, with the outside edge of the set hammer directly over the edge of the anvil. While hammering in this position keep the work turning continually.

To draw stock means to make it longer and thinner by hammering. A piece to be drawn out is usually laid across the horn of the anvil while being struck with the hammer. The metal is then spread in only one direction in place of being spread in every direction, as it would be if laid on the anvil face. To draw the work, heat it to as high a temperature as it will stand without throwing sparks and burning. The fuller may be used for drawing metal in place of laying the work over the horn of the anvil.

When drawing round stock, it should be first drawn out square, and when almost down to size it may be rounded. When pointing stock, the same rule of first drawing out square applies.

Upsetting means to make a piece shorter in length and greater in thickness or width, or both shorter and thicker. To upset short pieces, heat to a bright red at the place to be upset, then stand on end on the anvil face and hammer directly down on top until of the right form. Longer pieces may be swung against the anvil or placed upright on a heavy piece of metal lying on the floor or that is sunk into the floor. While standing on this heavy piece the metal may be upset by striking down on the end with a heavy hammer or the sledge. If a bend appears while upsetting, it should be straightened by hammering back into shape on the anvil face.

Light blows affect the metal for only a short distance from the point of striking, but heavy blows tend to swell the metal more equally through its entire length. In driving rivets that should fill the holes, heavy blows should be struck, but to shape the end of a rivet or to make a head on a rod, light blows should be used.

The part of the piece that is heated most will upset the most.

To punch a hole through metal, use a tool steel punch with its end slightly tapering to a size a little smaller than the hole to be punched. The end of the punch must be square across and never pointed or rounded.

First drive the punch part way through from one side and then turn the work over. When you turn it over, notice where the bulge appears and in that way locate the hole and drive the punch through from the second side. This makes a cleaner and more even hole than to drive completely through from one side. When the punch is driven in from the second side, the place to be punched through should be laid over the spud hole in the tail of the anvil and the piece driven out of the work.

Work when hot is larger than it will be after cooling. This must be remembered when fitting parts or trouble will result. A two-foot bar of steel will be 1/4 inch longer when red hot than when cold.

The temperatures of iron correspond to the following colors:

Dullest red seen in the dark... 878°
Dullest red seen in daylight... 887°
Dull red...................... 1100°
Full red...................... 1370°
Light red..................... 1550°
Orange........................ 1650°
Light orange.................. 1725°
Yellow........................ 1825°
Light yellow.................. 1950°

Bending Pipes and Tubes. — It is difficult to make bends or curves in pipes and tubing without leaving a noticeable bulge at some point of the work. Seamless steel tubing may be handled without very great danger of this trouble if care is used, but iron pipe, having a seam running lengthwise, must be given special attention to avoid opening the seam.

Bends may be made without kinking if the tube or pipe is brought to a full red heat all the way around its circumference and at the place where the bend is desired. Hold the cool portion solidly in a vise and, by taking hold of the free end, bend very slowly and with a steady pull. The pipe must be kept at full red heat with the flames from one or more torches and must not be hammered to produce the bend. If a sufficient purchase cannot be secured on the free end by the hand, insert a piece of rod or a smaller pipe into the opening.

While making the bend, should small bulges appear, they may be hammered back into shape before proceeding with the work.

Tubing or pipes may be bent while being held between two flat metal surfaces while at a bright red heat. The metal plates at each side of the work prevent bulging.

Another method by which tubing may be bent consists of filling completely with tightly packed sand and fitting a solid cap or plug at each end.

Thin brass tubing may be filled with melted resin and may be bent after the resin cools. To remove the resin it is necessary to heat the tube, allowing it to run out.

Large jobs of bending should be handled in special pipe bending machines in which the work is forced through formed rolls which prevent its bulging.

WELDING

Welding with the heat of a blacksmith forge fire, or a coal or illuminating gas fire, can only be performed with iron and steel because of the low heat which is not localized as with the oxy-acetylene and electric processes. Iron to be welded in this manner is heated until it reaches the temperature indicated by an orange color, not white, as is often stated, this orange color being slightly above 3600 degrees Fahrenheit. Steel is usually welded at a bright red heat because of the danger of oxidizing or burning the metal if the temperature is carried above this point.

The Fire. — If made in a forge, the fire should be built from good smithing coal or, better still, from coke. Gas fires are, of course, produced by suitable burners and require no special preparation except adjustment of the heat to the proper degree for the size and thickness of the metal being welded so that it will not be burned.

A coal fire used for ordinary forging operations should not be used for welding because of the impurities it contains. A fresh fire should be built with a rather deep bed of coal, four to eight inches being about right for work ordinarily met with. The fire should be kept burning until the coal around the edges has been thoroughly coked and a sufficient quantity of fuel should be on and around the fire so that no fresh coal will have to be added while working.

After the coking process has progressed sufficiently, the edges should be packed down and the fire made as small as possible while still surrounding the ends to be joined. The fire should not be altered by poking it while the metal is being heated. The best form of fire to use is one having rather high banks of coked coal on each side of the mass, leaving an opening or channel from end to end. This will allow the added fuel to be brought down on top of the fire with a small amount of disturbance.

Preparing to Weld.—If the operator is not familiar with the metal to be handled, it is best to secure a test piece if at all possible and try heating it and joining the ends. Various grades of iron and steel call for different methods of handling and for different degrees of heat, the proper method and temperature being determined best by actual test under the hammer.

The form of the pieces also has a great deal to do with their handling, especially in the case of a more or less inexperienced workman. If the pieces are at all irregular in shape, the motions should be gone through with before the metal is heated and the best positions on the anvil as well as in the fire determined with regard to the convenience of the workman and speed of handling the work after being brought to a welding temperature. Unnatural positions at the anvil should be avoided as good work is most difficult of performance under these conditions.

Scarfing.—While there are many forms of welds, depending on the relative shape of the pieces to be joined, the portions that are to meet and form one piece are always shaped in the same general way, this shape being called a "scarf." The end of a piece of work, when scarfed, is tapered off on one side so that the extremity comes to a rather sharp edge. The other side of the piece is left flat and a continuation in the same straight plane with its side of the whole piece of work. The end is then in the form of a bevel or mitre joint (Figure 50).

[Illustration: Figure 50.—Scarfing Ends of Work Ready for Welding]

Scarfing may be produced in any one of several ways. The usual method is to bring the ends to a forging heat, at which time they are upset to give a larger body of metal at the ends to be joined. This body of metal is then hammered down to the taper on one side, the length of the tapered portion being about one and a half times the thickness of the whole piece being handled. Each piece should be given this shape before proceeding farther.

The scarf may be produced by filing, sawing or chiseling the ends, although this is not good practice because it is then impossible to give the desired upset and additional metal for the weld. This added thickness is called for by the fact that the metal burns away to a certain extent or turns to scale, which is removed before welding.

When the two ends have been given this shape they should not fit as closely together as might be expected, but should touch only at the center of the area to be joined (Figure 51). That is to say, the surface of the beveled portion should bulge in the middle or should be convex in shape so that the edges are separated by a little distance when the pieces are laid together with the bevels toward each other. This is done so that the scale which is formed on the metal by the heat of the fire can have a chance to escape from the interior of the weld as the two parts are forced together.

[Illustration: Figure 51. — Proper Shape of Scarfed Ends]

If the scarf were to be formed with one or more of the edges touching each other at the same time or before the centers did so, the scale would be imprisoned within the body of the weld and would cause the finished work to be weak, while possibly giving a satisfactory appearance from the outside.

Fluxes. — In order to assist in removing the scale and other impurities and to make the welding surfaces as clean as possible while being joined, various fluxing materials are used as in other methods of welding.

For welding iron, a flux of white sand is usually used, this material being placed on the metal after it has been brought to a red heat in the fire. Steel is welded with dry borax powder, this flux being applied at the same time as the iron flux just mentioned. Borax may also be used for iron welding and a mixture of borax with steel borings may also be used for either class of work. Mixtures of sal ammoniac with borax have been successfully used, the proportions being about four parts of borax to one of sal ammoniac. Various prepared fluxing powders are on the market for this work, practically all of them producing satisfactory results.

After the metal has been in the fire long enough to reach a red heat, it is removed temporarily and, if small enough in size, the ends are dipped into a box of flux. If the pieces are large, they may simply be pulled to the edge of the fire and the flux then sprinkled on the portions to be joined. A greater quantity of flux is required in forge welding than in electric or oxy-acetylene processes because of the losses in the fire. After the powder has been applied to the surfaces, the work is returned to the fire and heated to the welding temperature.

Heating the Work. — After being scarfed, the two pieces to be welded are placed in the fire and brought to the correct temperature. This temperature can only be recognized by experiment and experience. The metal must be just below that point at which small sparks begin to be thrown out of the fire and naturally this is a hard point to distinguish. At the welding heat the metal is almost ready to flow and is about the consistency of putty. Against the background of the fire and coal the color appears to be a cream or very light yellow and the work feels soft as it is handled.

It is absolutely necessary that both parts be heated uniformly and so that they reach the welding temperature at the same time. For this reason they should be as close together in the fire as possible and side by side. When removed to be hammered together, time is saved if they are picked up in such a way that when laid together naturally the beveled surfaces come together. This makes it necessary that the workman remember whether the scarfed side is up or down, and to assist in this it is a good thing to mark the scarfed side with chalk or in some other noticeable manner, so that no mistake will be made in the hurry of placing the work on the anvil.

The common practice in heating allows the temperature to rise until the small white sparks are seen to come from the fire. Any heating above this point will surely result in burning that will ruin the iron or steel being handled. The best welding heat can be discerned by the appearance of the metal and its color after experience has been gained with this particular material. Test welds can be made and then broken, if possible, so that the strength gained through different degrees of heat can be known before attempting more important work.

Welding.—When the work has reached the welding temperature after having been replaced in the fire with the flux applied, the two parts are quickly tapped to remove the loose scale from their surfaces. They are then immediately laid across the top of the anvil, being placed in a diagonal position if both pieces are straight. The lower piece is rested on the anvil first with the scarf turned up and ready to receive the top piece in the position desired. The second piece must be laid in exactly the position it is to finally occupy because the two parts will stick together as soon as they touch and they cannot well be moved after having once been allowed to come in contact with each other. This part of the work must be done without any unnecessary loss of time because the comparatively low heat at which the parts weld allows them to cool below the working temperature in a few seconds.

The greatest difficulty will be experienced in withdrawing the metal from the fire before it becomes burned and in getting it joined before it cools below this critical point. The beveled edges of the scarf are, of course, the first parts to cool and the weld must be made before they reach a point at which they will not join, or else the work will be defective in appearance and in fact.

If the parts being handled are of such a shape that there is danger of bending a portion back of the weld, this part may be cooled by quickly dipping it into water before laying the work on the anvil to be joined.

The workman uses a heavy hand hammer in making the joint, and his helper, if one is employed, uses a sledge. With the two parts of the work in place on the anvil, the workman strikes several light blows, the first ones being at a point directly over the center of the weld, so that the joint will start from this point and be worked toward the edges. After the pieces have united the helper strikes alternate blows with his sledge, always striking in exactly the same place as the last stroke of the workman. The hammer blows are carried nearer and nearer to the edges of the weld and are made steadily heavier as the work progresses.

The aim during the first part of the operation should be to make a perfect joint, with every part of the surfaces united, and too much attention should not be paid to appearance, at least not enough to take any chance with the strength of the work.

It will be found, after completion of the weld, that there has been a loss in length equal to one-half the thickness of the metal being welded. This loss is occasioned by the burned metal and the scale which has been formed.

Finishing the Weld.—If it is possible to do so, the material should be hammered into the shape that it should remain with the same heat that was used for welding. It will usually be found, however, that the metal has cooled below the point at which it can be worked to advantage. It should then be replaced in the fire and brought back to a forging heat.

[Illustration: Figure 52.—Upsetting and Scarfing the End of a Rod]

While shaping the work at this forging heat every part that has been at a red heat should be hammered with uniformly light and even blows as it cools. This restores the grain and strength of the iron or steel to a great extent and makes the unavoidable weakness as small as possible

Forms of Welds.—The simplest of all welds is that called a "lap weld." This is made between the ends of two pieces of equal size and similar form by scarfing them as described and then laying one on top of the other while they are hammered together.

A butt weld (Figure 52) is made between the ends of two pieces of shaft or other bar shapes by upsetting the ends so that they have a considerable flare and shaping the face of the end so that it is slightly higher in the center than around the edges, this being done to make the centers come together first. The pieces are heated and pushed into contact, after which the hammering is done as with any other weld.

[Illustration: Figure 53.—Scarfing for a T Weld]

A form similar to the butt weld in some ways is used for joining the end of a bar to a flat surface and is called a jump weld. The bar is shaped in the same way as for a butt weld. The flat plate may be left as it is, but if possible a depression should be made at the point where the shaft is to be placed. With the two parts heated as usual, the bar is dropped into position and hammered from above. As soon as the center of the weld has been made perfect, the joint may be finished with a fuller driven all the way around the edge of the joint.

When it is required to join a bar to another bar or to the edge of any piece at right angles the work is called a "T" weld from its shape when complete (Figure 53). The end of the bar is scarfed as described and the point of the other bar or piece where the weld is to be made is hammered so that it tapers to a thin edge like one-half of a circular depression. The pieces are then laid together and hammered as for a lap weld.

The ends of heavy bar shapes are often joined with a "V," or cleft, weld. One bar end is shaped so that it is tapering on both sides and comes to a broad edge like the end of a chisel. The other bar is heated to a forging temperature and then slit open in a lengthwise direction so that the V-shaped opening which is formed will just receive the pointed edge of the first piece. With the work at welding heat, the two parts are driven together by hammering on the rear ends and the hammering then continues as with a lap weld, except that the work is turned over to complete both sides of the joint.

[Illustration: Figure 54.-Splitting Ends to Be Welded in Thin Work]

The forms so far described all require that the pieces be laid together in the proper position after removal from the fire, and this always causes a slight loss of time and a consequent lowering of the temperature. With very light stock, this fall of temperature would be so rapid that the weld would be unsuccessful, and in this case the "lock" weld is resorted to. The ends of the two pieces to be joined are split for some distance back, and one-half of each end is bent up and the other half down (Figure 54). The two are then pushed together and placed in the fire in this position. When the welding heat is reached, it is only necessary to take the work out of the fire and hammer the parts together, inasmuch as they are already in the correct position.

Other forms of welds in which the parts are too small to retain their heat, can be made by first riveting them together or cutting them so that they can be temporarily fastened in any convenient way when first placed in the fire.

CHAPTER VIII

SOLDERING, BRAZING AND THERMIT WELDING

SOLDERING

Common solder is an alloy of one-half lead with one-half tin, and is called "half and half." Hard solder is made with two-thirds tin and one-third lead. These alloys, when heated, are used to join surfaces of the same or dissimilar metals such as copper, brass, lead, galvanized iron, zinc, tinned plate, etc. These metals are easily joined, but the action of solder with iron, steel and aluminum is not so satisfactory and requires greater care and skill.

The solder is caused to make a perfect union with the surfaces treated with the help of heat from a soldering iron. The soldering iron is made from a piece of copper, pointed at one end and with the other end attached to an iron rod and wooden handle. A flux is used to remove impurities from the joint and allow the solder to secure a firm union with the metal surface. The iron, and in many cases the work, is heated with a gasoline blow torch, a small gas furnace, an electric heater or an acetylene and air torch.

The gasoline torch which is most commonly used should be filled two-thirds full of gasoline through the hole in the bottom, which is closed by a screw plug. After working the small hand pump for 10 to 20 strokes, hold the palm of your hand over the end of the large iron tube on top of the torch and open the gasoline needle valve about a half turn. Hold the torch so that the liquid runs down into the cup below the tube and fills it. Shut the gasoline needle valve, wipe the hands dry, and set fire to the fuel in the cup. Just as the gasoline fire goes out, open the gasoline needle valve about a half turn and hold a lighted match at the end of the iron tube to ignite the mixture of vaporized gasoline and air. Open or close the needle valve to secure a flame about 4 inches long.

On top of the iron tube from which the flame issues there is a rest for supporting the soldering iron with the copper part in the flame. Place the iron in the flame and allow it to remain until the copper becomes very hot, not quite red, but almost so.

A new soldering iron or one that has been misused will have to be "tinned" before using. To do this, take the iron from the fire while very hot and rub the tip on some flux or dip it into soldering acid. Then rub the tip of the iron on a stick of solder or rub the solder on the iron. If the solder melts off the stick without coating the end of the iron, allow a few drops to fall on a piece of tin plate, then nil the end of the iron on the tin plate with considerable force. Alternately rub the iron on the solder and dip into flux until the tip has a coating of bright solder for about half an inch from the end. If the iron is in very bad shape, it may be necessary to scrape or file the end before dipping in the flux for the first time. After the end of the iron is tinned in this way, replace it on the rest of the torch so that the tinned point is not directly in the flame, turning the flame down to accomplish this.

Flux.—The commonest flux, which is called "soldering acid," is made by placing pieces of zinc in muriatic (hydrochloric) acid contained in a heavy glass or porcelain dish. There will be bubbles and considerable heat evolved and zinc should be added until this action ceases and the zinc remains in the liquid, which is now chloride of zinc.

This soldering acid may be used on any metal to be soldered by applying with a brush or swab. For electrical work, this acid should be made neutral by the addition of one part ammonia and one part water to each three parts of the acid. This neutralized flux will not corrode metal as will the ordinary acid.

Powdered resin makes a good flux for lead, tin plate, galvanized iron and aluminum. Tallow, olive oil, beeswax and vaseline are also used for this purpose. Muriatic acid may be used for zinc or galvanized iron without the addition of the zinc, as described in making zinc chloride. The addition of two heaping teaspoonfuls of sal ammoniac to each pint of the chloride of zinc is sometimes found to improve its action.

Soldering Metal Parts. — All surfaces to be joined should be fitted to each other as accurately as possible and then thoroughly cleaned with a file, emery cloth, scratch bush or by dipping in lye. Work may be cleaned by dipping it into nitric acid which has been diluted with an equal volume of water. The work should be heated as hot as possible without danger of melting, as this causes the solder to flow better and secure a much better hold on the surfaces. Hard solder gives better results than half and half, but is more difficult to work. It is very important that the soldering iron be kept at a high heat during all work, otherwise the solder will only stick to the surfaces and will not join with them.

Sweating is a form of soldering in which the surfaces of the work are first covered with a thin layer of solder by rubbing them with the hot iron after it has been dipped in or touched to the soldering stick. These surfaces are then placed in contact and heated to a point at which the solder melts and unites. Sweating is much to be preferred to ordinary soldering where the form of the work permits it. This is the only method which should ever be used when a fitting is to be placed over the end of a length of tube.

Soldering Holes. — Clean the surfaces for some distance around the hole until they are bright, and apply flux while holding the hot iron near the hole. Touch the tip of the iron to some solder until the solder is picked up on the iron, and then place this solder, which was just picked up, around the edge of the hole. It will leave the soldering iron and stick to the metal. Keep adding solder in this way until the hole has been closed up by working from the edges and building toward the center. After the hole is closed, apply more flux to the job and smooth over with the hot iron until there are no rough spots. Should the solder refuse to flow smoothly, the iron is not hot enough.

Soldering Seams. — Clean back from the seam or split for at least half an inch all around and then build up the solder in the same way as was done with the hole. After closing the opening, apply more flux to the work and run the hot iron lengthwise to smooth the job.

Soldering Wires. — Clean all insulation from the ends to be soldered and scrape the ends bright. Lay the ends parallel to each other and, starting at the middle of the cleaned portion, wrap the ends around each other, one being wrapped to the right, the other to the left. Hold the hot iron under the twisted joint and apply flux to the wire. Then dip the iron in the solder and apply to the twisted portion until the spaces between the wires are filled with solder. Finish by smoothing the joint and cleaning away all excess metal by rubbing the hot iron lengthwise. The joint should now be covered with a layer of rubber tape and this covered with a layer of ordinary friction tape.

Steel and Iron.—Steel surfaces should be cleaned, then covered with clear muriatic acid. While the acid is on the metal, rub with a stick of zinc and then tin the surfaces with the hot iron as directed. Cast iron should be cleaned and dipped in strong lye to remove grease. Wash the lye away with clean water and cover with muriatic acid as with steel. Then rub with a piece of zinc and tin the surfaces by using resin as a flux.

It is very difficult to solder aluminum with ordinary solder. A special aluminum solder should be secured, which is easily applied and makes a strong joint. Zinc or phosphor tin may be used in place of ordinary solder to tin the surfaces or to fill small holes or cracks. The aluminum must be thoroughly heated before attempting to solder and the flux may be either resin or soldering acid. The aluminum must be thoroughly cleaned with dilute nitric acid and kept hot while the solder is applied by forcible rubbing with the hot iron.

BRAZING

This is a process for joining metal parts, very similar to soldering, except that brass is used to make the joint in place of the lead and zinc alloys which form solder. Brazing must not be attempted on metals whose melting point is less than that of sheet brass.

Two pieces of brass to be brazed together are heated to a temperature at which the brass used in the process will melt and flow between the surfaces. The brass amalgamates with the surfaces and makes a very strong and perfect joint, which is far superior to any form of soldering where the work allows this process to be used, and in many cases is the equal of welding for the particular field in which it applies.

Brazing Heat and Tools.—The metal commonly used for brazing will melt at heats between 1350° and 1650° Fahrenheit. To bring the parts to this temperature, various methods are in use, using solid, liquid or gaseous fuels. While brazing may be accomplished with the fire of the blacksmith forge, this method is seldom satisfactory because of the difficulty of making a sufficiently clean fire with smithing coal, and it should not be used when anything else is available. Large jobs of brazing may be handled with a charcoal fire built in the forge, as this fuel produces a very satisfactory and clean fire. The only objection is in the difficulty of confining the heat to the desired parts of the work.

The most satisfactory fire is that from a fuel gas torch built for this work. These torches are simply forms of Bunsen burners, mixing the proper quantity of air with the gas to bring about a perfect combustion. Hose lines lead to the mixing tube of the gas torch, one line carrying the gas and the other air under a moderate pressure. The air line is often dispensed with, allowing the gas to draw air into the burner on the injector principle, much the same as with illuminating gas burners for use with incandescent mantles. Valves are provided with which the operator may regulate the amount of both gas and air, and ordinarily the quality and intensity of the flame.

When gas is not available, recourse may be had to the gasoline torch made for brazing. This torch is built in the same way as the small portable gasoline torches for soldering operations, with the exception that two regulating needle valves are incorporated in place of only one.

The torches are carried on a framework, which also supports the work being handled. Fuel is forced to the torch from a large tank of gasoline into which air pressure is pumped by hand. The torches are regulated to give the desired flame by means of the needle valves in much the same way as with any other form of pressure torch using liquid fuel.

Another very satisfactory form of torch for brazing is the acetylene-air combination described in the chapter on welding instruments. This torch gives the correct degree of heat and may be regulated to give a clean and easily controlled flame.

Regardless of the source of heat, the fire or flame must be adjusted so that no soot is deposited on the metal surfaces of the work. This can only be accomplished by supplying the exact amounts of gas and air that will produce a complete burning of the fuel. With the brazing torches in common use two heads are furnished, being supplied from the same source of fuel, but with separate regulating devices. The torches are adjustably mounted in such a way that the flames may be directed toward each other, heating two sides of the work at the same time and allowing the pieces to be completely surrounded with the flame.

Except for the source of heat, but one tool is required for ordinary brazing operations, this being a spatula formed by flattening one end of a quarter-inch steel rod. The spatula is used for placing the brazing metal on the work and for handling the flux that is required in this work as in all other similar operations.

Spelter.—The metal that is melted into the joint is called spelter. While this name originally applied to but one particular grade or composition of metal, common use has extended the meaning until it is generally applied to all grades.

Spelter is variously composed of alloys containing copper, zinc, tin and antimony, the mixture employed depending on the work to be done. The different grades are of varying hardness, the harder kinds melting at higher temperatures than the soft ones and producing a stronger joint when used. The reason for not using hard spelter in all cases is the increased difficulty of working it and the fact that its melting point is so near to some of the metals brazed that there is great danger of melting the work as well as the spelter.

The hardest grade of spelter is made from three-fourths copper with one-fourth zinc and is used for working on malleable and cast iron and for steel.

This hard spelter melts at about 1650° and is correspondingly difficult to handle.

A spelter suitable for working with copper is made from equal parts of copper and zinc, melting at about 1400° Fahrenheit, 500°

below the melting point of the copper itself. A still softer brazing metal is composed of half copper, three-eighths zinc and one-eighth tin. This grade is used for fastening brass to iron and copper and for working with large pieces of brass to brass. For brazing thin sheet brass and light brass castings, a metal is used which contains two-thirds tin and one-third antimony. The low melting point of this last composition makes it very easy to work with and the danger of melting the work is very slight. However, as might be expected, a comparatively weak joint is secured, which will not stand any great strain.

All of the above brazing metals are used in powder form so that they may be applied with the spatula where the joint is exposed on the outside of the work. In case it is necessary to braze on the inside of a tube or any deep recess, the spelter may be placed on a flat rod long enough to reach to the farthest point. By distributing the spelter at the proper points along the rod it may be placed at the right points by turning the rod over after inserting into the recess.

Flux.—In order to remove the oxides produced under brazing heat and to allow the brazing metal to flow freely into place, a flux of some kind must be used. The commonest flux is simply a pure calcined borax powder, that is, a borax powder that has been heated until practically all the water has been driven off.

Calcined borax may also be mixed with about 15 per cent of sal ammoniac to make a satisfactory fluxing powder. It is absolutely necessary to use flux of some kind and a part of whatever is used should be made into a paste with water so that it can be applied to the joint to be brazed before heating. The remainder of the powder should be kept dry for use during the operation and after the heat has been applied.

Preparing the Work.—The surfaces to be brazed are first thoroughly cleaned with files, emery cloth or sand paper. If the work is greasy, it should be dipped into a bath of lye or hot soda water so that all trace of oil is removed. The parts are then placed in the relation to each other that they are to occupy when the work has been completed. The edges to be joined should make a secure and tight fit, and should match each other at all points so that the smallest possible space is left between them. This fit should not be so tight that it is necessary to force the work into place, neither should it be loose enough to allow any considerable space between the surfaces. The molten spelter will penetrate between surfaces that water will flow between when the work and spelter have both been brought to the proper heat. It is, of course, necessary that the two parts have a sufficient number of points of contact so that they will remain in the proper relative position.

The work is placed on the surface of the brazing table in such a position that the flame from the torches will strike the parts to be heated, and with the joint in such a position that the melted spelter will flow down through it and fill every possible part of the space between the surfaces under the action of gravity. That means that the edge of the joint must be uppermost and the crack to be filled must not lie horizontal, but at the greatest slant possible. Better than any degree of slant would be to have the line of the joint vertical.

The work is braced up or clamped in the proper position before commencing to braze, and it is best to place fire brick in such positions that it will be impossible for cooling draughts of air to reach the heated metal should the flame be removed temporarily during the process. In case there is a large body of iron, steel or copper to be handled, it is often advisable to place charcoal around the work, igniting this with the flame of the torch before starting to braze so that the metal will be maintained at the correct heat without depending entirely on the torch.

When handling brass pieces having thin sections there is danger of melting the brass and causing it to flow away from under the flame, with the result that the work is ruined. If, in the judgment of the workman, this may happen with the particular job in hand, it is well to build up a mould of fire clay back of the thin parts or preferably back of the whole piece, so that the metal will have the necessary support. This mould may be made by mixing the fire clay into a stiff paste with water and then packing it against the piece to be supported tightly enough so that the form will be retained even if the metal softens.

Brazing.—With the work in place, it should be well covered with the paste of flux and water, then heated until this flux boils up and runs over the surfaces. Spelter is then placed in such a position that it will run into the joint and the heat is continued or increased until the spelter melts and flows in between the two surfaces. The flame should surround the work during the heating so that outside air is excluded as far as is possible to prevent excessive oxidization.

When handling brass or copper, the flame should not be directed so that its center strikes the metal squarely, but so that it glances from one side or the other. Directing the flame straight against the work is often the cause of melting the pieces before the operation is completed. When brazing two different metals, the flame should play only on the one that melts at the higher temperature, the lower melting part receiving its heat from the other. This avoids the danger of melting one before the other reaches the brazing point.

The heat should be continued only long enough to cause the spelter to flow into place and no longer. Prolonged heating of any metal can do nothing but oxidize and weaken it, and this practice should be avoided as much as possible. If the spelter melts into small globules in place of flowing, it may be caused to spread and run into the joint by lightly tapping the work. More dry flux may be added with the spatula if the tapping does not produce the desired result.

Excessive use of flux, especially toward the end of the work, will result in a very hard surface on all the work, a surface which will be extremely difficult to finish properly. This trouble will be present to a certain extent anyway, but it may be lessened by a vigorous scraping with a wire brush just as soon as the work is removed from the fire. If allowed to cool before cleaning, the final appearance will not be as good as with the surplus metal and scale removed immediately upon completing the job.

After the work has been cleaned with the brush it may be allowed to cool and finished to the desired shape, size and surface by filing and polishing. When filed, a very thin line of brass should appear where the crack was at the beginning of the work. If it is desired to avoid a square shoulder and fill in an angle joint to make it rounding, the filling is best accomplished by winding a coil of very thin brass wire around the part of the work that projects and then causing this to flow itself or else allow the spelter to fill the spaces between the layers of wire. Copper wire may also be used for this purpose, the spaces being filled with melted spelter.

THERMIT WELDING

The process of welding which makes use of the great heat produced by oxygen combining with aluminum is known as the Thermit process and was perfected by Dr. Hans Goldschmidt. The process, which is controlled by the Goldschmidt Thermit Company, makes use of a mixture of finely powdered aluminum with an oxide of iron called by the trade name, Thermit.

The reaction is started with a special ignition powder, such as barium superoxide and aluminum, and the oxygen from the iron oxide combining with the aluminum, producing a mass of superheated steel at about 5000 degrees Fahrenheit. After the reaction, which takes from. 30 seconds to a minute, the molten metal is drawn from the crucible on to the surfaces to be joined. Its extreme heat fuses the metal and a perfect joint is the result. This process is suited for welding iron or steel parts of comparatively large size.

Preparation. — The parts to be joined are thoroughly cleaned on the surfaces and for several inches back from the joint, after which they are supported in place. The surfaces between which the metal will flow are separated from 1/4 to 1 inch, depending on the size of the parts, but cutting or drilling part of the metal away. After this separation is made for allowing the entrance of new metal, the effects of contraction of the molten steel are cared for by preheating adjacent parts or by forcing the ends apart with wedges and jacks. The amount of this last separation must be determined by the shape and proportions of the parts in the same way as would be done for any other class of welding which heats the parts to a melting point.

Yellow wax, which has been warmed until plastic, is then placed around the joint to form a collar, the wax completely filling the space between the ends and being provided with vent holes by imbedding a piece of stout cord, which is pulled out after the wax cools.

A retaining mould (Figure 55) made from sheet steel or fire brick is then placed around the parts. This mould is then filled with a mixture of one part fire clay, one part ground fire brick and one part fire sand. These materials are well mixed and moistened with enough water so that they will pack. This mixture is then placed in the mould, filling the space between the walls and the wax, and is packed hard with a rammer so that the material forms a wall several inches thick between any point of the mould and the wax. The mixture must be placed in the mould in small quantities and packed tight as the filling progresses.

[Illustration: Figure 55.—Thermit Mould Construction]

Three or more openings are provided through this moulding material by the insertion of wood or pipe forms. One of these openings will lead from the lowest point of the wax pattern and is used for the introduction of the preheating flame. Another opening leads from the top of the mould into this preheating gate, opening into the preheating gate at a point about one inch from the wax pattern. Openings, called risers, are then provided from each of the high points of the wax pattern to the top of the mould, these risers ending at the top in a shallow basin. The molten metal comes up into these risers and cares for contraction of the casting, as well as avoiding defects in the collar of the weld. After the moulding material is well packed, these gate patterns are tapped lightly and withdrawn, except in the case of the metal pipes which are placed at points at which it would be impossible to withdraw a pattern.

Preheating.—The ends to be welded are brought to a bright red heat by introducing the flame from a torch through the preheating gate. The torch must use either gasoline or kerosene, and not crude oil, as the crude oil deposits too much carbon on the parts. Preheating of other adjacent parts to care for contraction is done at this time by an additional torch burner.

The heating flame is started gently at first and gradually increased. The wax will melt and may be allowed to run out of the preheating gate by removing the flame at intervals for a few seconds. The heat is continued until the mould is thoroughly dried and the parts to be joined are brought to the red heat required. This leaves a mould just the shape of the wax pattern.

The heating gate should then be plugged with a sand core, iron plug or piece of fitted fire brick, and backed up with several shovels full of the moulding mixture, well packed.

[Illustration: Figure 56.—Thermit Crucible Plug. *A*, Hard burn magnesia stone; *B*, Magnesia thimble; *C*, Refractory sand; *D*, Metal disc; *E*, Asbestos washer; *F*, Tapping pin]

Thermit Metal. — The reaction takes place in a special crucible lined with magnesia tar, which is baked at a red heat until the tar is driven off and the magnesia left. This lining should last from twelve to fifteen reactions. This magnesia lining ends at the bottom of the crucible in a ring of magnesia stone and this ring carries a magnesia thimble through which the molten steel passes on its way to the mould. It will usually be necessary to renew this thimble after each reaction. This lower opening is closed before filling the crucible with thermit by means of a small disc or iron carrying a stem, which is called a tapping pin (Figure 56). This pin, *F*, is placed in the thimble with the stem extending down through the opening and exposing about two inches. The top of this pin is covered with an asbestos, washer, *E*, then with another iron disc. *D*, and finally with a layer of refractory sand. The crucible is tapped by knocking the stem of the pin upwards with a spade or piece of flat iron about four feet long.

The charge of thermit is added by placing a few handfuls over the refractory sand and then pouring in the balance required. The amount of thermit required is calculated from the wax used. The wax is weighed before and after filling *the entire space that the thermit will occupy*. This does not mean only the wax collar, but the space of the mould with all gates filled with wax. The number of pounds of wax required for this filling multiplied by 25 will give the number of pounds of thermit to be used. To this quantity of thermit should be added I per cent of pure manganese, 1 per cent nickel thermit and 15 per cent of steel punchings.

It is necessary, when more than 10 pounds of thermit will be used, to mix steel punchings not exceeding 3/8 inch diameter by 1/8 inch thick with the powder in order to sufficiently retard the intensity of the reaction.

Half a teaspoonful of ignition powder is placed on top of the thermit charge and ignited with a storm match or piece of red hot iron. The cover should be immediately closed on the top of the crucible and the operator should get away to a safe distance because of the metal that may be thrown out of the crucible.

After allowing about 30 seconds to a minute for the reaction to take place and the slag to rise to the top of the crucible, the tapping pin is struck from below and the molten metal allowed to run into the mould. The mould should be allowed to remain in place as long as possible, preferably over night, so as to anneal the steel in the weld, but in no case should it be disturbed for several hours after pouring. After removing the mould, drill through the metal left in the riser and gates and knock these sections off. No part of the collar should be removed unless absolutely necessary.

CHAPTER IX

OXYGEN PROCESS FOR REMOVAL OF CARBON

Until recently the methods used for removing carbon deposits from gas engine cylinders were very impractical and unsatisfactory. The job meant dismantling the motor, tearing out all parts, and scraping the pistons and cylinder walls by hand.

The work was never done thoroughly. It required hours of time to do it, and then there was always the danger of injuring the inside of the cylinders.

These methods have been to a large extent superseded by the use of oxygen under pressure. The various devices that are being manufactured are known as carbon removers, decarbonizers, etc., and large numbers of them are in use in the automobile and gasoline traction motor industry.

Outfit.—The oxygen carbon cleaner consists of a high pressure oxygen cylinder with automatic reducing valve, usually constructed on the diaphragm principle, thus assuring positive regulation of pressure. This valve is fitted with a pressure gauge, rubber hose, decarbonizing torch with shut off and flexible tube for insertion into the chamber from which the carbon is to be removed.

There should also be an asbestos swab for swabbing out the inside of the cylinder or other chamber with kerosene previous to starting the operation. The action consists in simply burning the carbon to a fine dust in the presence of the stream of oxygen, this dust being then blown out.

Operation.—The following are instructions for operating the cleaner:—

(1) Close valve in gasoline supply line and start the motor, letting it run until the gasoline is exhausted.

(2) If the cylinders be T or L head, remove either the inlet or the exhaust valve cap, or a spark plug if the cap is tight. If the cylinders have overhead valves, remove a spark plug. If any spark plug is then remaining in the cylinder it should be removed and an old one or an iron pipe plug substituted.

(3) Raise the piston of the cylinder first to be cleaned to the top of the compression stroke and continue this from cylinder to cylinder as the work progresses.

(4) In motors where carbon has been burned hard, the cylinder interior should then be swabbed with kerosene before proceeding. Work the swab, saturated with kerosene, around the inside of the cylinder until all the carbon has been moistened with the oil. This same swab may be used to ignite the gas in the cylinder in place of using a match or taper.

(5) Make all connections to the oxygen cylinder.

(6) Insert the torch nozzle in the cylinder, open the torch valve gradually and regulate to about two lbs. pressure. Manipulate the nozzle inside the cylinder and light a match or other flame at the opening so that the carbon starts to burn. Cover the various points within the cylinder and when there is no further burning the carbon has been removed. The regulating and oxygen tank valves are operated in exactly the same way as for welding as previously explained.

It should be carefully noted that when the piston is up, ready to start the operation, both valves must be closed. There will be a considerable display of sparks while this operation is taking place, but they will not set fire to the grease and oil. Care should be used to see that no gasoline is about.

INDEX

Acetylene
 filtering
 generators
 in tanks
 piping
 properties of
 purification of
Acetylene-air torches
Air
 oxygen from
Alloys
 table of
Alloy steel
Aluminum
 alloys
 welding
Annealing
Anvil
Arc welding, electric
 machines
Asbestos, use of, in welding

Babbitt Bending pipes and tubes Bessemer steel Beveling Brass welding Brazing electric heat and tools spelter Bronze welding Butt welding

Calcium carbide Carbide storage of, Fire Underwriters' Rules to water generator Carbon removal by oxygen process Case hardening steel Cast iron welding Champfering Charging generator Chlorate of potash oxygen Conductivity of metals Copper alloys welding Crucible steel Cutting, oxy-acetylene torches

Dissolved acetylene

Electric arc welding
Electric welding
  troubles and remedies
Expansion of metals

Flame, welding
Fluxes
  for brazing
  for soldering
Forge
  fire
  practice
  tools
  tuvere construction of
  welding
  welding preparation
  welds, forms of
Forging

Gas holders
Gases, heating power of
Generator, acetylene
  carbide to water
  construction
Generator
  location of
  operation and care of
  overheating
  requirements
  water to carbide
German silver
Gloves
Goggles

Hand forging
Hardening steel
Heat treatment of steel
Hildebrandt process

Hose

Injectors, adjuster
Iron
  cast
  grades of
  malleable cast
  wrought

  Jump weld

Lap welding
Lead
Linde process
Liquid air oxygen

Magnalium
Malleable iron
  welding
Melting points of metals
Metal alloys, table of
Metals
  characteristics of
  conductivity of
  expansion of
  heat treatment of
  melting points of
  tensile strength of
  weight of

Nickel
Nozzle sizes, torch

Open hearth steel
Oxy-acetylene cutting
  welding practice
Oxygen
  cylinders

weight of

Pipes, bending
Platinum
Preheating

Removal of carbon by oxygen process
Resistance method of electric welding
Restoration of steel
Rods, welding

Safety devices
Scarfing
Solder
Soldering
  flux
  holes
  seams
  steel and iron
  wires
Spelter
Spot welding
Steel
  alloys
  Bessemer
  crucible
  heat treatment of
  open hearth
  restoration of
  tensile strength of
  welding
Strength of metals

Tank valves
Tapering
Tables of welding information
Tempering steel
Thermit metal
  preheating

  preparation
  welding
Tin
Torch
  acetylene-air
  care
  construction
  cutting
  high pressure
  low pressure
  medium pressure
  nozzles
  practice

Valves, regulating
  tank

Water
  to carbide generator
Welding aluminum
  brass
  bronze
  butt
  cast iron
  copper
  electric
  electric arc
  flame
  forge
  information and tables
  instruments
  lap
  malleable iron
  materials
  practice, oxy-acetylene
  rods
  spot
  steel
  table

thermit
  torches
  various metals
  wrought iron
Wrought iron
  welding

   Zinc

www.ingramcontent.com/pod-product-compliance
Lightning Source LLC
Chambersburg PA
CBHW030031250526
45464CB000026B/1868